POLLEN

POLLEN

THE HIDDEN SEXUALITY OF FLOWERS

Rob Kesseler & Madeline Harley

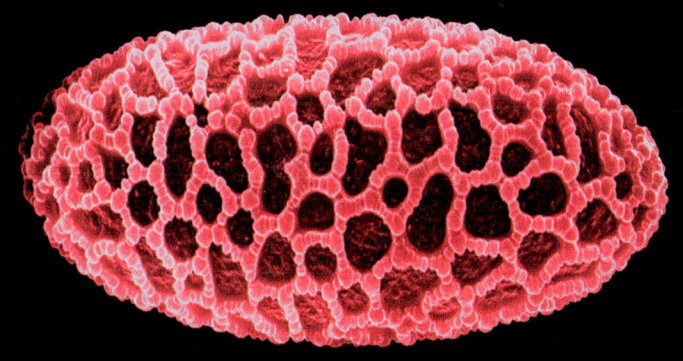

PAPADAKIS PUBLISHER

Acknowledgments

This book would not have been possible without the help and support of many people. We are grateful to our publisher Andreas Papadakis for help in the difficult task of combining science and art in the development of an important book that serves both disciplines, and to Alexandra Papadakis for a design concept that does justice to both. We would especially like to thank NESTA (National Endowment for Science, Technology and Art) for their generous financial support and guidance throughout, and especially for Rob Kesseler's Fellowship at Kew; and Peter Crane, Director of the Royal Botanic Gardens, Kew, for letting an artist into the laboratories. We are indebted to many members of staff at Kew who have been generous with their time and expertise in answering our numerous questions, and giving their support and encouragement to this project, in particular: Simon Owens (Keeper, Herbarium), Mike Bennett (Keeper, Jodrell Laboratory), Paula Rudall (Head of Micromorphology), David Cooke (Temperate House), Tom Cope (Herbarium), Hannah Rogers and Ali Cuthbert (Press Office), Gina Fullerlove (Publications), Laura Giuffrida (Manager, Exhibitions and Live Interpretation, HPE), Tony Kirkham (Head, Arboretum & Horticultural Services), Paola Magris, Marilyn Ward, Sam Cox, and James Kay (Illustrations Collections). At NESTA we would like especially to thank Alex Barclay, Joe Meaney, and Sara Macnee. At Central St Martin's School of Art & Design, we would like to thank Jonathan Barratt (Dean of the School of Graphics and Industrial Design), Kathryn Hearn (Course Director, Ceramic Design), Stuart Evans (Senior Lecturer, Research). Among all the other people whose help and support have helped us see this project through we thank Stephen Blackmore (Regius Keeper, Edinburgh Botanic Garden), Basil and Anette Harley (Harley Books), Robert Hewison, Paul Holt (Senior Project Manager, Samphire Hoe), Martin Kemp (Oxford University), Kathy Meek, Maria Suarez-Cervera (University of Barcelona), Adam Sutherland (Director, Grizedale Arts), Robert Woof (Director, Wordsworth Trust), Tim Green (BBC), Roger Huyton (BBC). Lastly a very special thank you to Agalis Manessi and Marco Kesseler for their endless patience and encouragement.

Rob Kesseler & Madeline Harley

Editorial and design director: Alexandra Papadakis

First published in 2004 in Great Britain by
PAPADAKIS PUBLISHER
An imprint of New Architecture Group Ltd
16 Grosvenor Place, London SW1X 7HH
www.papadakis.net

ISBN 1 901092 67 4
New Edition 2006

Picture Credits

Royal Botanic Gardens, Kew; Natural History Museum Library; National Museum of Photography, Film & Television/Science & Society Picture Library; Anthony Cragg courtesy of The Lisson Gallery; University of Jena; McGraw Hill; Alexandra Papadakis; Claire Waring; Almqvist & Wiksell

We gratefully acknowledge the granting of permission to use these images. Every reasonable attempt has been made to identify and contact copyright holders. Any errors or omissions are inadvertent and will be corrected in subsequent editions.

The Royal Botanic Gardens, Kew

The Royal Botanic Gardens, Kew is both a world renowned scientific organization and a major visitor attraction. Over a million people a year explore this World Heritage Site; 132 hectares of imposing gardens with six glasshouses, a lake and ponds, a Victorian picture gallery and a museum exhibition. Behind the scenes are internationally important collections of living plants, a world-class herbarium, and global research programmes that inform biodiversity and plant conservation work throughout the world. About 150,000 children visit Kew each year, and students of horticulture graduate from a renowned on-site school. Kew is also responsible for the National Trust's most visited site, Wakehurst Place in Sussex, home to the Millennium Seed Bank Project.

CONTENTS

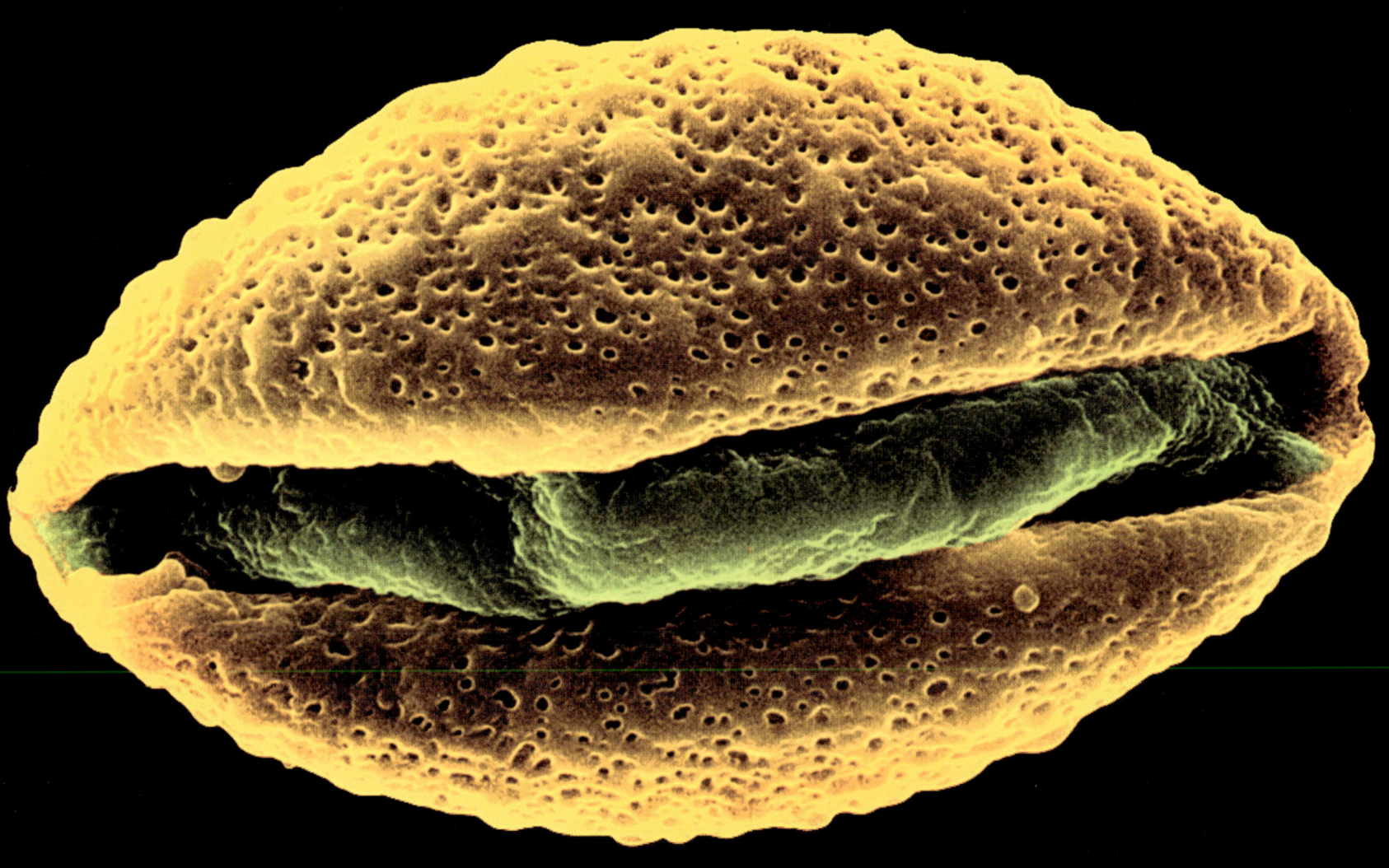

Liriodendron tulipifera – Tulip Tree (Magnoliaceae) – pollen
grain, natural state – dehydrated [SEM x 1500]
opposite: *Liriodendron tulipifera* – Tulip Tree (Magnoliaceae) –
an open flower showing mature stamens – most of the pollen
has been released from the anthers. Dauntsey Park, Wiltshire

We dedicate this book to Nehemiah Grew (1641–1712)
whose remarkable and profound observations on pollen,
"particles of prolifick virtue", have been a source of wonder and delight.

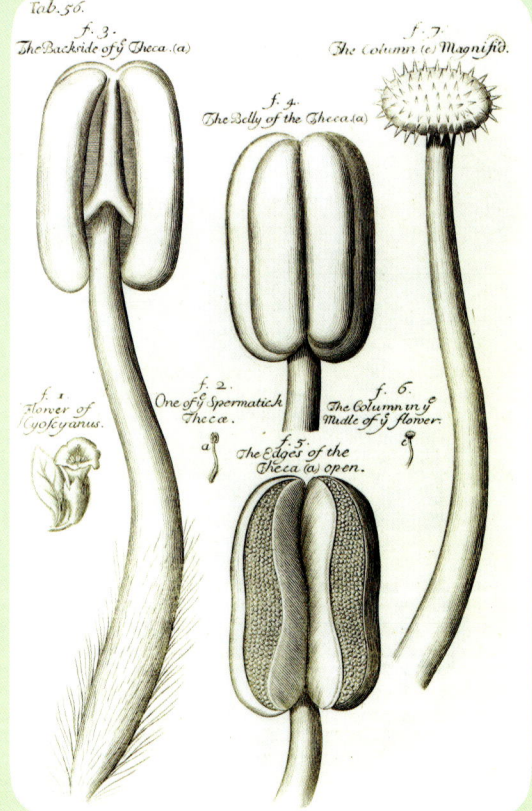

"The **particles** of these powders, though like those of meal or other dust, they appear not easily to have any regular shape; yet upon strict observation, especially with the assistance of an indifferent glass, it doth appear, that there are a congeries [collections, masses], usually, of so many globes or globulets; sometimes of other figures, but always regular." … "Of the secondary use hereof, I have spoken in the first book; and particularly, of the globulets or small particles within the thecae of the seed-like attire [stamen], and upon the blades [pistils] of the florid, I have conjectured, that they are the body which bees gather and carry upon their thighs, and is commonly called their bread. For the wax they carry in little flakes in their chaps [cheeks]: but the bread is a kind of powder; yet somewhat moist, as are the said little particles of the attire." … "But the primary and chief use of the attire is such, as hath respect to the plant itself; and so appears to be very great and necessary. Because, even those plants which have no flower or foliature, are yet some way or other attir'd; either with the seminiform [a stamen with the filament base attached to floral receptacle], or the florid attire [stamens of tubular flowers with the filaments attached to the floral tube]. So that it seems to perform its service to the seed, as the foliature to the fruit." … "And as the young and early attire before it opens, answers to the menses in the femal [sic]: so it is probable, that afterward when it opens or cracks, it performs the office of the male. This is hinted from the shape of the parts. For in the florid attire, the blade doth not unaptly resemble a small penis, with the sheath upon it, as its præputinum [foreskin]. And in the seed-like attire, the several thecae, are so many little testicles. And the globulets and other small particles upon the blade or penis, and in the thecae, are as the vegetable sperme. Which, so soon as the penis is exerted, or the testicles come to break, falls down upon the seed-case or womb, and so touches it with **prolifick virtue**."

Nehemiah Grew: *The Anatomy of Flowers, Prosecuted with the bare Eye, and with the Microscope*, 1682

EXPLANATORY NOTES

It was never our intention to write a text book; rather a book for everyone to enjoy and discover the extraordinary life and form of pollen grains. Nevertheless, to explain what pollen is, what it looks like, and what it does, we need to introduce and use a number of technical words in the scientific section. For some the botanical terminology and/or the pollen terminology will be commonplace, but we are hoping to attract a far wider audience than the botanically informed. Therefore, for those readers who may, at times, feel verbally challenged we have indicated the first entry of each specialist word with a pair of single apostrophes, for example 'pollenkitt'. Usually we explain the specialist words where they first occur, but to avoid over-explanation, and spoil the flow of the narrative, we have also provided a comprehensive glossary at the end of the book.

BOTANICAL LATIN: The notes compiled here draw exclusively on the most authoritative, human and widely used account of botanical Latin ever published. Accepted as a channel of communication so distinct from classical Latin in spirit and structure as to require independent treatment, the book, *Botanical Latin*, is the work of William Stearn, and it grew from the seed of an idea planted in the young Stearn's mind from an imperfect attempt to help an Indian botany student to translate his description of a new species into Latin for formal publication in the Botanical Journal of the Linnean Society. The experience made the youthful Stearn realise that a knowledge of classical Latin was of little practical use in botanical Latin names and plant descriptions – "the realm of literature which a knowledge of botanical Latin opens to botanists is a strange barbarous place for classicists". Stearn started his "huge self-inflicted task" during the Second World War as a way of filling in time as he watched the sky for planes from a Royal Air Force ambulance.

William (or Willie as he was affectionately known) Stearn 'left no stone unturned' in searching out, correlating and analysing Latin botanical descriptions and glossaries in order to distil out of them all a clear account of the rules governing the language of botanical Latin. After the War, beginning with his wartime notebooks "off and on among other tasks", it was to take Stearn twenty years before his magnum opus was published in 1966. It sold out in ten months! Since then three further editions have been published (1973, 1983, 1992).

LATIN BOTANICAL NAMES: There is a protocol, to which we adhere in this book, when citing the Latin names of plants. For example, Pussy Willow is the 'common name' for *Salix caprea* a member of the Willow family – Salicaceae. All formally named species of plants have a genus name and a species name – a binomial: *Salix* is the genus, and always begins with a capital letter, and *caprea* is the species name and, nowadays, always begins with a lower case letter – in the past a species name might be given a capital letter if the plant was named for a person or a place. The genus name may be used meaningfully without the species name, because all generic names are unique – for example, *Salix* or *Salix* sp. – from which we can tell that the author refers to a species (sp.) of *Salix*, but the species is not given in the text or is unknown. However, we cannot use the species name (which is an adjectival appellation or 'epithet') meaningfully without including the genus because there may be other genera which have a species with the same epithet – for example, *Myosotis arvensis* (Common Forget-me-not) and *Sonchus arvensis* (Field Milk Thistle). If we are making a list of species of *Salix* to be more concise we would write *Salix* in full only for the first species in the list and then abbreviate to *'S.': Salix alba, S. caprea, S. fragilis*, etc. Note also that the Latin name is given in italics; this is standard practice for genus and species names.

WHY LATIN NAMES? Plants and animals are given Latin names to make life easier, not more difficult. Latin is the international language of the natural sciences, and Latin names and terminology are understood by naturalists worldwide. For example, while an English, a French and a German botanist will immediately understand each other if talking about *Papaver rhoeas*, they might be some while in realising they were all discussing the same species of plant if they used their nationally accepted common names: Field Poppy, Pavot Rouge and Klatschmohn, respectively.

It is not uncommon to find that a name used by florists, or at a Garden Centre, is not botanically correct and/or up to date. There are some examples in this book. At the time of writing there is a member of the Gentian family (Gentianaceae), *Lisianthus*, which is commonly sold as a cut flower in supermarkets. The botanically correct name is *Eustoma grandiflorum*, and the authority (Rafin.) Shinners. Formerly the plant was *Lisianthius grandiflorus* Aubl. (the third 'i' in *Lisianthius* is missing from the label on the supermarket bouquets). The geraniums we grow in pots and borders in the summer, although in the same family as the true *Geranium*, are actually cultivars of *Pelargonium* species, while most of the *Geranium* species and cultivars that we grow in our gardens are clump-forming hardy perennials, although some of our British wild *Geraniums* are annuals, for example the delightful, if somewhat pungent, Herb Robert (*Geranium robertianum L.*).

AUTHORITIES FOR LATIN NAMES: All formally recognised plant species are given a 'binomial' Latin name the first time they are described formally in a scientific publication. If the name is later revised, or changed in any way, this change must be formally published as well. The names are given by botanists specialising in plant taxonomy (the classification of organisms). There are many authorities for plant names. "L." is in fact a scientifically recognised abbreviation for Carl Linnaeus (1707-1778), the Swedish naturalist who introduced the Latin binomial system for the naming of plants and animals. He named many species of plants and animals, particularly, but not by any means exclusively, in Northwest Europe. Note in the *Index of Plants Illustrated* (p. 263) at the end of this book how many times the authority for a species is 'L.' In a scientific text the first time a species is mentioned by the botanist the authority for the name should be cited alongside the name, for example *Salix caprea* L. In this book the authorities are not cited in the text, as this would interrupt the flow of the text, especially for the non-botanist. Instead we have provided the authorities for names in the *List of Plants Illustrated* near the end of the book. The reason for citing the authority is to provide a reference point for that name – the genus and species can then be checked to ascertain whether this is the correct and current name for the plant. Botanical and zoological nomenclature is a complex and highly specialised subject, bound by many rules and regulations, and governed by *The International Code for Botanical Nomenclature.*

BOTANICAL LATIN: THE GREEK ELEMENT: It is to William Stearn's *Botanical Latin* (Fourth Edition – pp. 252-274) that we turn for an explanation of why there are so many Greek words in botanical Latin: "Although Latin is the official language for the scientific names of plants, many such names are really Greek in origin. The cause is two-fold. As E.L. Greene noted: 'Pliny, the supreme Latin writer about plants, in translating Theophrastan texts by the hundred into Latin for Roman readers, made use of familiar Latin names in place of the Greek names when there were such, ...' For many plants, however, there were no Latin names available. Pliny overcame this difficulty by transliterating the Greek name into Roman characters, the termination being sometimes changed by him or the not always competent clerks and scribes working hurriedly on his vast compilation, in order to conform to Latin usage ... Linnaeus listed many others [such names] ... and himself drew upon ancient names to designate new genera."

"There are, however, many botanical names which, although compounded of Greek words, formed no part of ancient Greek. Such names are continually being introduced. This is partly because the apt Latin word has been used already, but chiefly because Greek is a rich flexible language in which pleasing compounds are readily made."

FAMILY NAMES OF PLANTS: These usually end with –aceae. In some cases family names end with –ae: Labiatae, Compositae, Leguminosae and Cruciferae are among the best known examples. These names are conserved because they have been in use by botanists and horticulturalists for a long time and are not only well known they also belong to very large families of plants! All these families have an alternative conserved name ending with the more usual –aceae: Lamiaceae, Asteraceae, Fabaceae, and Brassicaceae, respectively.

COMMON (VERNACULAR) NAMES OF PLANTS: There is also a protocol for common names of plants – if we are referring to generalised plant groups, for example, potatoes, tomatoes, cucumbers, roses, we use lower case for the initial letter of the word. If we are referring to the common name of a particular species of plant, for example, Dandelion, we use upper case for the initial letter. If the common name has more than one word we use upper case for each word: Pussy Willow, or Dog Violet, but not if the words are hyphenated: Forget-me-not, Lords-and-ladies. Common names are not italicised. There can also be confusion with the same common name, in different regions: for example, in Scotland the Bluebell is *Campanula rotundifolia* L. (family Campanulaceae) while in England it is *Endymion non-scriptus* (L.) Garcke (family Liliaceae). To pursue this subject in greater depth the following two books are recommended (full details are given in the *Bibliography*): *Botanical Latin* by William Stearn and *Biological Nomenclature* by Charles Jeffrey.

BIBLIOGRAPHY: For those interested in exploring some of the topics described and discussed in the book, bibliographies of selected references to pollen and pollination, and to art and artists are provided.

ILLUSTRATIONS: With the important exception of selected images by other botanists and artists, mainly historical, most of the images included here are the original work of the authors. Pollen images are either from specially collected and prepared material, or from past research studies. Flower images, in particular close up details of flowers, were photographed with a Nikon D100 using 60mm micro nikkor and 35-105 macro nikkor lenses. Pollen grain preparation techniques are described in the text. For light microscopy (LM) the prepared pollen grains were photographed using a Nikon Optiphot, fitted with a 100x oil immersion objective; for scanning electron microscopy (SEM) the pollen was examined and photographed using a Hitachi S2400 SEM, and for transmission electron microscopy (TEM) the pollen grains were embedded in specially prepared resin and thin sectioned using a Reichert Ultracut fitted with a Diatome diamond knife. The resultant thin sections were stained in an LKB ultrostainer, and subsequently examined and photographed using a Hitachi H300 TEM. Light microscope images were output on 35mm high resolution colour, or black and white film; SEM and TEM images were output as high resolution black and white negatives. Subsequently the images were digitally scanned, and some of the SEM images were manipulated to introduce colour. The selective use of colour was driven by a number of factors: sometimes simply in response to the original flower or pollen colour; sometimes to draw out aspects of pollen structure or function, while other choices were purely intuitive responses, and wilfully unscientific.

The selection process for pollen images was clearly driven by the differing concerns of the scientist and the artist, the choices providing starting points for discussion and reflection on our respective practices. Many of the pollen grains were prepared to show them fully expanded, and cleaned, with ornamentation and other features clearly visible. Other specimens were transferred directly from fresh flowers, left to dry naturally, and then examined in the SEM in a 'warts and all' approach. Aesthetic considerations were given precedence over the more usual fact-seeking examination; collapsed or even imperfectly formed pollen grains, which might have been rejected in routine examination, were especially chosen for their sculptural forms.

MAGNIFICATION OF POLLEN IMAGES

Although the original magnifications at which the pollen grains were photographed are given, many of the images presented in this volume have been further enlarged.

ABBREVIATIONS USED IN THE TEXT AND FIGURE LEGENDS:

LM	= light microscopy
SEM	= scanning electron microscopy
TEM	= transmission electron microscopy
CPD	= critical point drying
cv.	= a garden cultivar, not a natural species
sp.	= species (singular)
spp.	= more than one species
sp. unk.	= species unknown

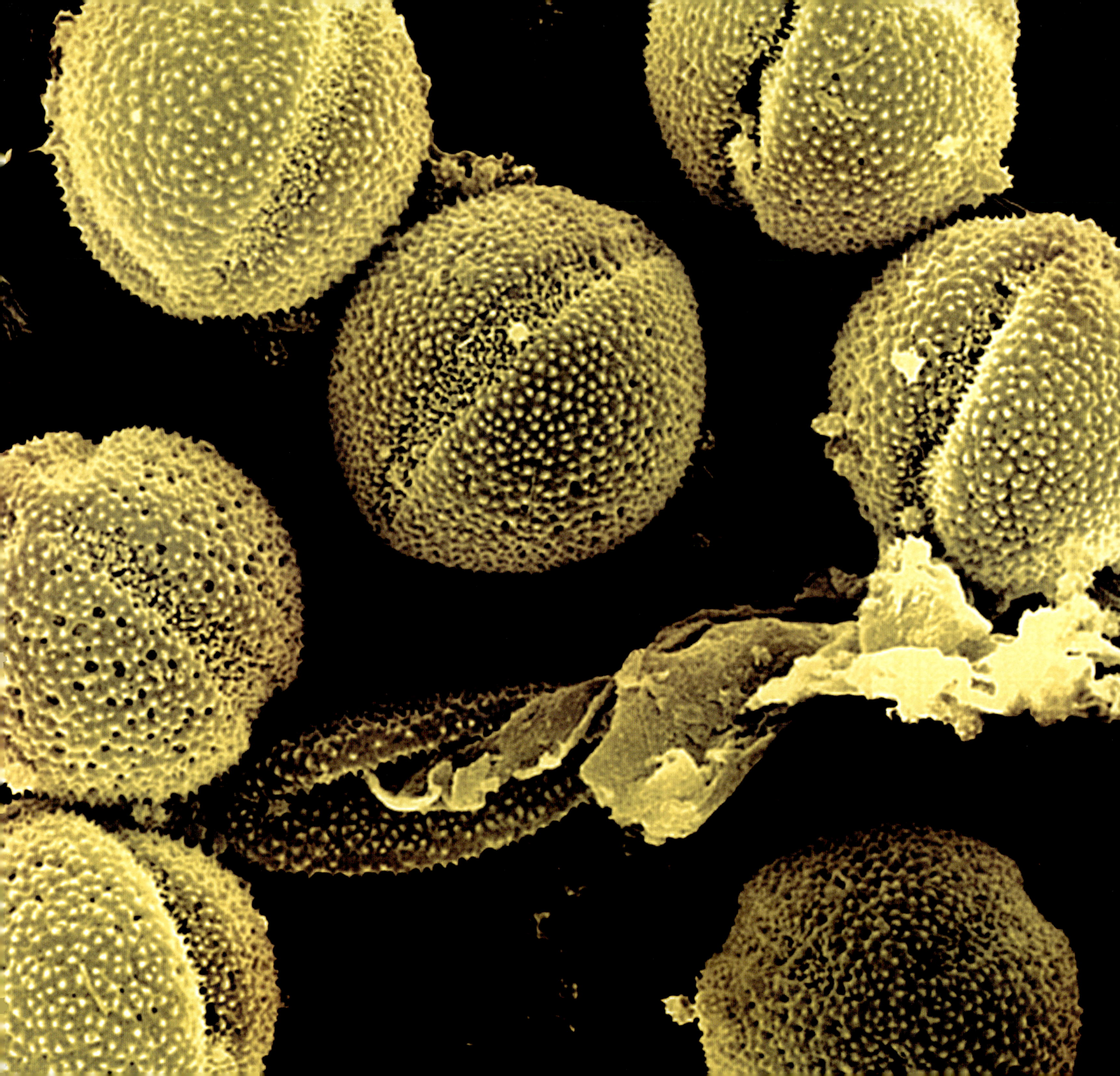

Meconopsis cambrica (L.) Vig. – Welsh Poppy (Papaveraceae) –
pollen grains, expanded natural condition, note coating of pollenkitt
[SEM x 1500]

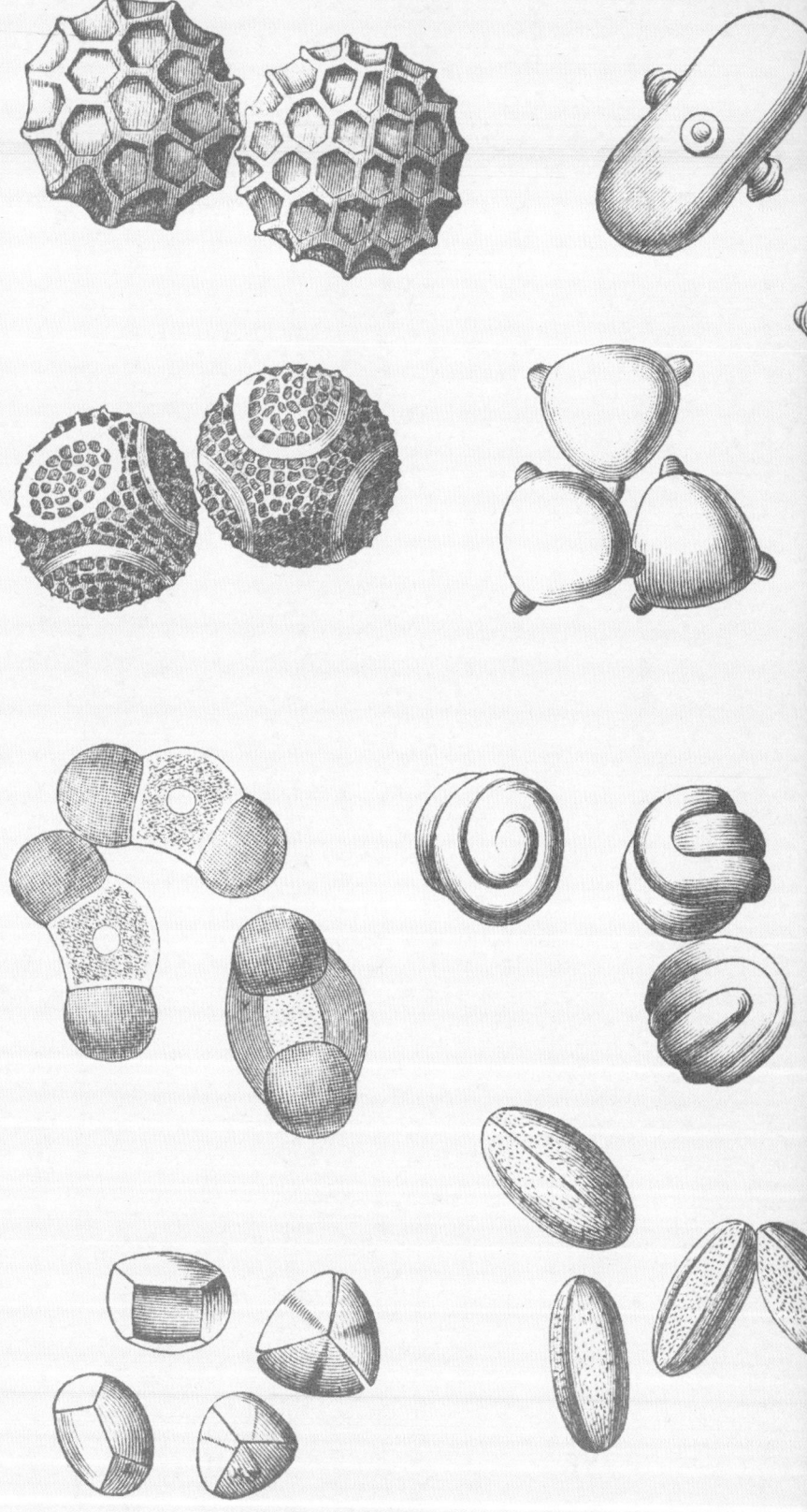

For almost 250 years the Royal Botanic Gardens, Kew have been associated with the finest in the artistic representation of botanical subjects. In the eighteenth century three generations of the Hanoverian Royal family – Princess Augusta, Queen Charlotte and Princess Elizabeth – were instructed in flower painting by Margaret Meen and Franz Bauer, two of the most renowned botanical painters of their day. They chose their subjects from the burgeoning plant collections gathered together at Kew. Later, in the nineteenth century, Walter Hood Fitch and his contemporaries captured the essence of exotic plants brought back by overseas collectors, from the giant amazonian waterlily to the bizarre *Welwitschia*. At the same time, Marianne North travelled the world in search of scenes and plant subjects that now grace the Gallery that carries her name. Throughout the twentieth century the tradition of botanical art at Kew has been maintained unbroken through the exquisite work of Harriet Thistleton-Dyer, Matilda Smith, Lilian Snelling, Margaret Mee, Margaret Stone, Stella Ross-Craig, Mary Grierson and many, many others. And today it continues, as Kew expands its unrivalled collection of botanical art through targeted acquisitions, and through the commission of new work for *Curtis's Botanical Magazine* and other publications. There are also a series of active training programmes that help develop the botanical artists of the future.

The centre of gravity of botanical art has always been the accurate representation of whole plants and flowers, which in turn has been inextricably linked to the scientific objectives of documenting plant diversity. However, in the context of Kew's broader scientific mission, there has always been a parallel strand of art that has focused on the microscopic structure of plants. Ever since the early work of Nehemiah Grew and others, both artists and scientists have been fascinated by the intricate structures revealed by microscopes.

Especially important in the development of art that represents botanical microstructure was Franz Bauer. Bauer, through his meticulous attention to detail, was among the very first to probe the diversity of form among pollen grains, thereby helping to provide the

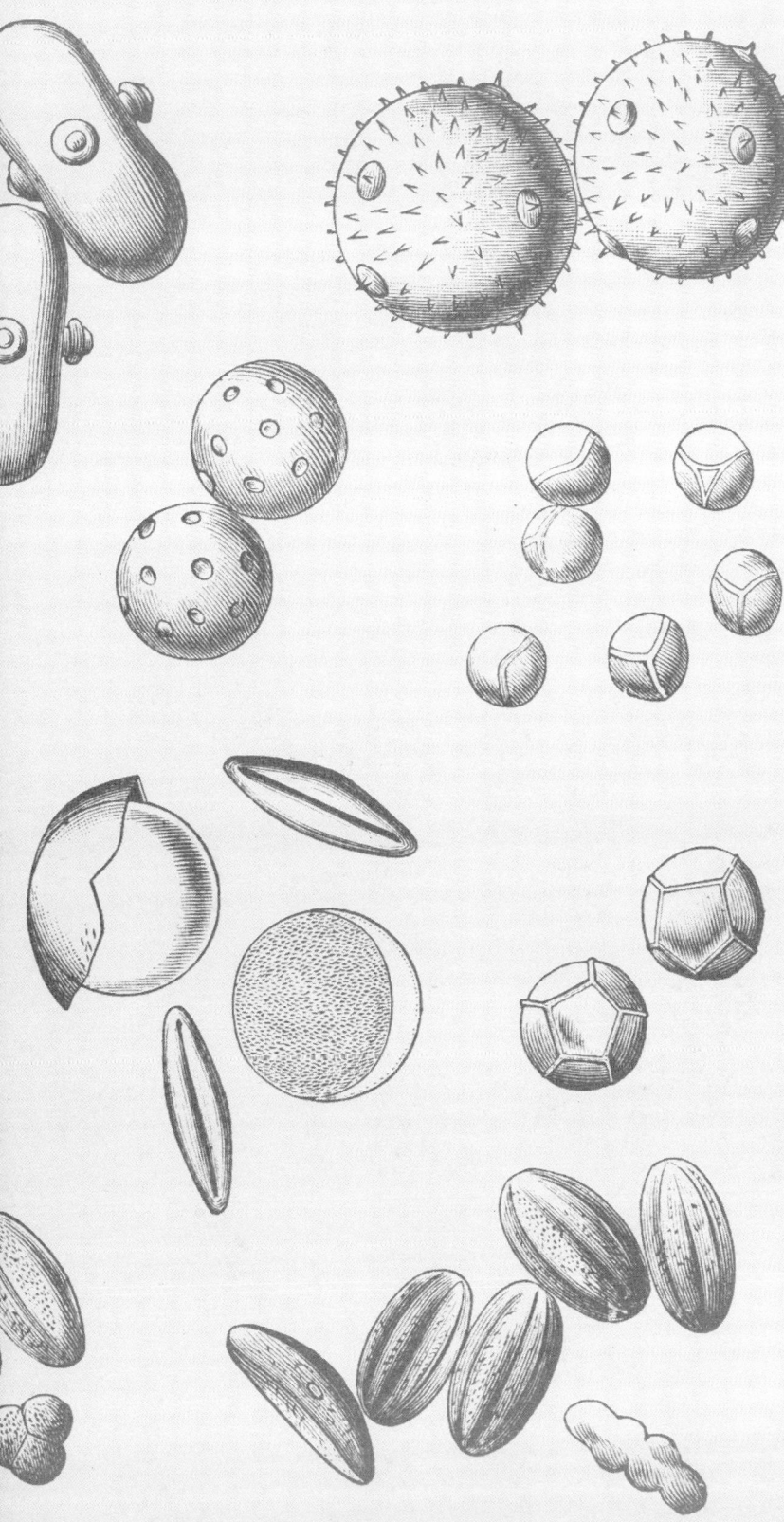

foundation for specialist and more intensive study of pollen in the nineteenth century. Today Bauer's work, and the other aspects of plant structure that have been revealed subsequently, are of continuing scientific interest. There is also new intellectual emphasis on searching for the secrets on how genetic information is translated into the intricacies of plant form.

The unique collaboration between a scientist and an artist represented by this book therefore extends a long tradition in which the exploration of the complexities of plant structure is intimately entwined with artistic attempts to analyse, understand and represent botanical form. It continues the exploration of pollen diversity begun by prominent scientists in the seventeenth and eighteenth centuries, and also illuminates the key function of pollen in the life cycle of plants. And by utilising scanning electron photomicroscopy this book takes us into worlds that even comparatively recently were still inaccessible. The resulting studies are inherently fascinating both as art, and for the insights they provide into how plants work. Madeline Harley and Rob Kesseler should be congratulated on producing a book that will delight, inspire and intrigue. I hope that through their dedication and creativity, you enjoy these new views of some of the hidden masterpieces in the world of plants.

Professor Sir Peter Crane

background: A.Kerner & F.W.Oliver, illustration of pollen grains from *The Natural History of Plants*, 1903.

THE ART AND SCIENCE OF POLLEN

ROB KESSELER & MADELINE HARLEY

T his book is the result of the shared fascination of an artist and a scientist with the perfect design of organisms too small to be seen without a microscope – pollen grains, which are enclosed beyond the accessible beauty of the flower until the moment of release, when they will be carried by wind, water or animal vectors to achieve their purpose, procreation. Pollen is ubiquitous; in childhood we all learn a little of plant reproduction and the role of the bee but few people are aware of the astonishing diversity of the structure of pollen grains, although these tiny and extraordinary forms have fascinated the scientifically curious since the seventeenth century.

Throughout history there have been polymath geniuses whose passion for understanding enabled them to traverse many disciplines, Leonardo da Vinci being the exemplar. In the seventeenth century came Robert Hooke, chemist, physicist and surveyor of the City of London, whose pioneering development of the compound microscope was to have such an impact on the scientific world. Printed in 1665, his seminal book *Micrographia* was a landmark in popular science publishing. Not only did Hooke describe in an accessible language his microscopic observations of anything from woven strands of silk to a flea but also he illustrated each specimen with graphic precision, giving the subjects an 'other worldly' appearance. We have become so used to the legacy of richly illustrated books which it engendered, that it is hard to imagine the sensation that *Micrographia* caused when it was first published. It pre-dated by almost a century the proliferation of popular illustrated books generated to satisfy the growing fascination across Europe for collecting and displaying 'nature's ornaments', with titles such as *Spectakulum Naturae &*

Artium, 1765, and *Amusemens Microscopiques, 1776.* The relationship between art and science has ebbed and flowed since Hooke's time. Goethe, who would send himself to sleep at night by visualising the developmental cycle of plants, had a less happy time. He was surprised and dismayed to find that his essay, *On the Metamorphosis of Plants*, although recognised thirty years later as a serious contribution to botany, was ignored at the time by botanists and public alike. He complained, "Nowhere would anyone grant that science and poetry can be united. People forgot that science had developed from poetry and they failed to take into consideration that a swing of the pendulum might benificently (sic) reunite the two, at a higher level and to mutual advantage."

After a period of separation, the cultures of science and art are currently enjoying a collaborative renaissance. This book is a testament to this new spirit of co-operation. The sophistication and quality of the images produced scientifically is such that they have a clarity and detail that may call into question the need for any artistic intervention at all. However, this would be to ignore the role of the artist in interpreting and translating new scientific imagery, acting as the conduit through which the cultural consequences of scientific discovery are developed. Contemporary audiences of all ages are showing a growing appetite for images of the natural world that not only evoke a sense of awe through their sheer magnificence but also offer the opportunity to learn more about the workings of life.

It is with great pleasure that we are now able to share the fruits of our science-art collaboration.

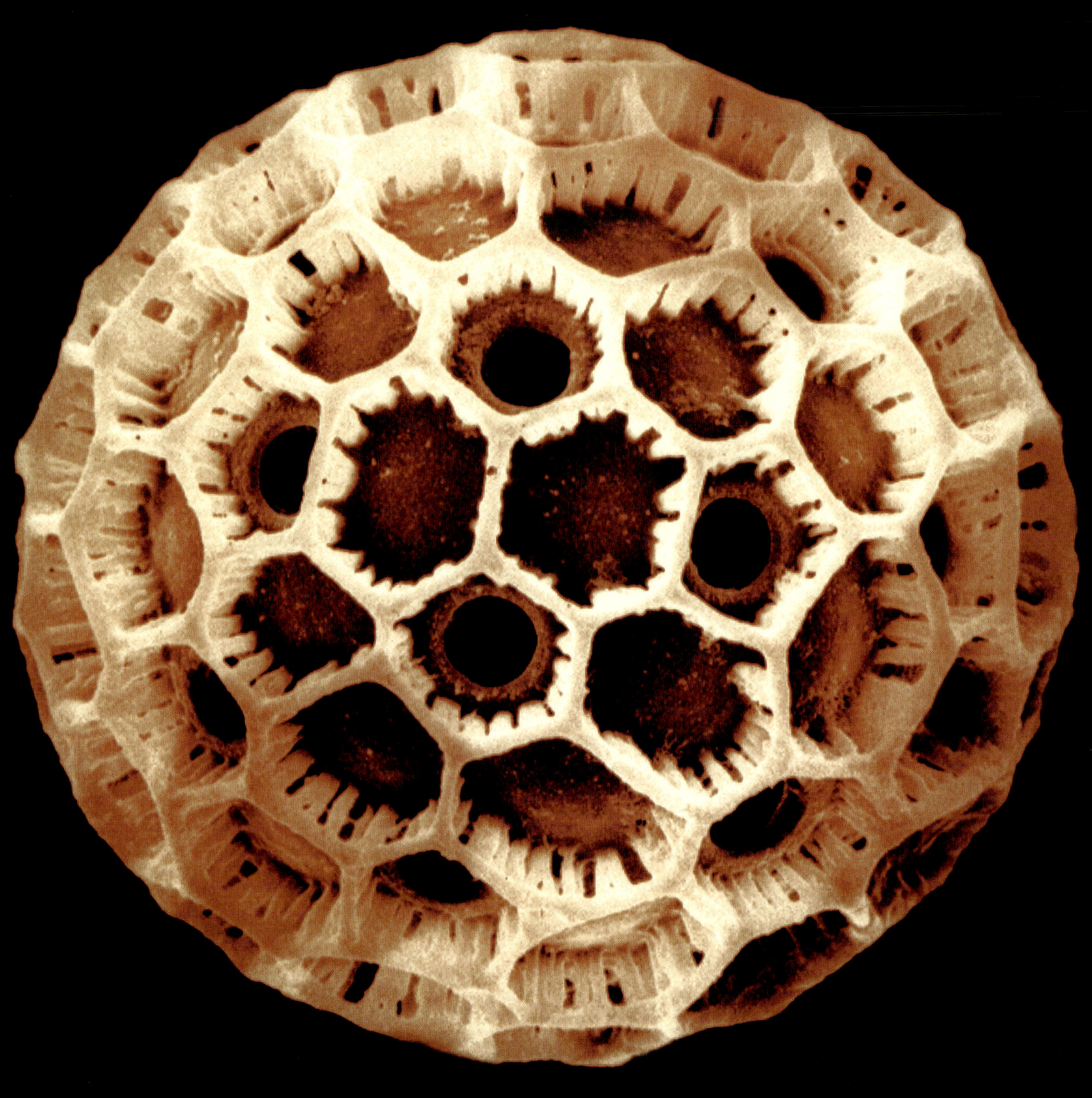

NO FLOWERS – NO POLLEN
NO POLLEN – NO FLOWERS

MADELINE HARLEY

Cobaea scandens – Cup and Saucer Vine (Polemoniaceae) –
pollen grain, expanded condition

This book is a celebration of the journey of mature pollen grains as they leave the anther on a quest to find a mate for the sperm they carry. It is not just the magical story of how pollen grains fertilise ovules so that plants develop seeds, it is also about the pollen grains themselves. They are, without doubt, among the most beautiful microscopic structures in nature. They are tiny – perfect masterpieces of natural architecture and structural engineering – and often breathtakingly beautiful. Their range of forms is extraordinary and, in isolation from the plant that produced them, we can observe their individual characteristics and either know, or narrow down the possible species from which they may have originated.

The word 'pollen' is Latin and means fine dust or flour; the use of the word in this context goes back to antiquity. Its first use as a scientific word to describe the male sperm carrying units of flowering plants is credited to Carl Linnaeus in *Sponsalia Plantarum* (The Betrothals of Plants) published in 1747. In the first edition of his book, *Philosophy of Botany* (1750(1)),[1] Linnaeus defines pollen: "Pollen est pulvis vegetabilium, appropriato liquore madefactus rumpendus, & substantium sensibus nudis imperscrutabile elastice explodens." (Pollen is the dust of vegetables, which will burst when moistened with the appropriate liquid, and propulsively explode a substance which is not discernable by the naked senses.)

Most of us are aware of pollen, mainly because it may stain our clothes or, more annoyingly and frequently, cause miserable allergenic reactions (hayfever). These irritants have little to do with the real reasons for the existence of pollen. Pollen grains are perfect and quite remarkable natural entities. Not only are they very small, almost imperceptible to the human eye – apart from appearing as dust – they are, more importantly, the extraordinarily structured containers for carrying the sperm cells of two major plant groups: flowering plants (angiosperms), and conifers and their relatives.

Lilium cv. – Florists' Lily (Liliaceae) – pollen grain, expanded condition [SEM x 1000]

opposite: *Lilium* cv. – Florists' Lily (Liliaceae) – close up of anthers

If we look at pollen grains down a microscope we enter a fantastic world where, although small is beautiful, use far outweighs ornament. The appearance of the tough outer casing of the pollen grain, which encloses the sperm cells, shows an amazing range of variation between different species of plants. These variations are frequently elaborate, often exquisitely so, and they are referred to as 'pollen types'. There are thousands of pollen types. Usually a plant species produces pollen of only one type. However, there are not as many pollen types as there are species of plants, and some species share a very similar pollen type with another species; particularly species that are closely related. Some pollen types are common to a number of plant 'families', and if the plant that produced the pollen is not to hand, it is difficult, even for an expert, to identify the plant that produced the pollen. Then there are families of plants, such as the grasses and bamboos (Poaceae), where the pollen is extraordinarily similar in all the species but, nevertheless, highly recognisable as grass pollen.

Most species of the cabbage and wallflower family, the Cruciferae share a similar pollen type; it is very different from the pollen of the Poaceae but again highly recognisable at family level. A plant family may have pollen that shares certain distinctive characteristics throughout the family but, in different species, shows variations of the basic pattern; the daisy family is a good example. Other families of plants, such as the Acanthaceae (*Acanthus*, 'Bear's Breeches', 'Black-eyed Susan') may have many types of pollen, although some or all of the types may be readily distinguished by an expert as being typical of a certain family.

FLOWER STRUCTURE

Before proceeding with our story a quick reminder of flower structure and terminology – perhaps redolent of half-remembered facts from our schooldays – might be helpful: a flower comprises a number of basic parts, but the variations in arrangement, structure and modification of these basic parts are

Lily flower cut away to show three of the six stamens, and the pistil

opposite: *Lilium* cv. – Florists' Lily (Liliaceae) – petals removed to show stamen arrangement

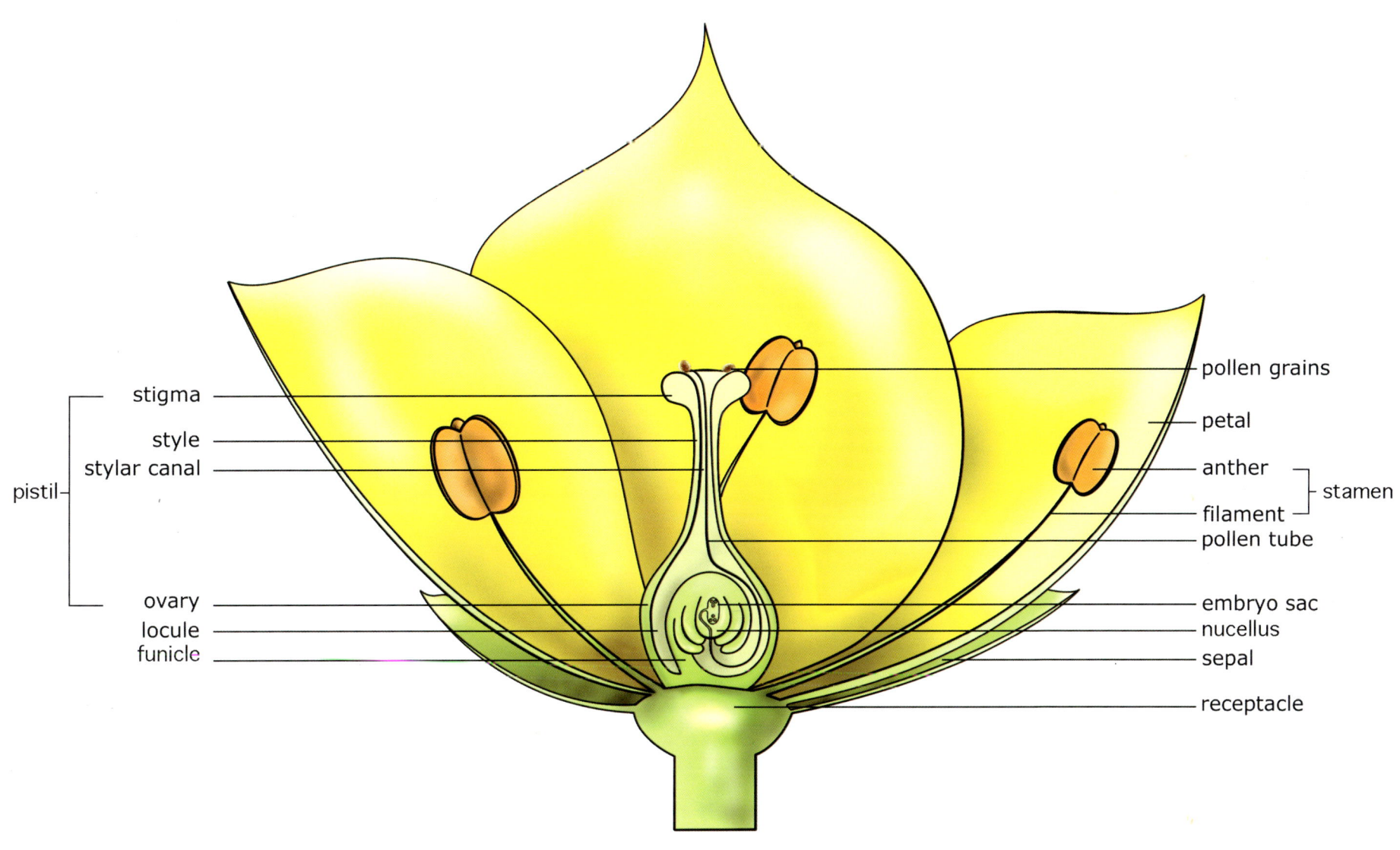

stigma

style

stylar canal

pistil

ovary

locule

funicle

pollen grains

petal

anther

filament

stamen

pollen tube

embryo sac

nucellus

sepal

receptacle

Parts of the flower

opposite: *Ranunculus acris* – Meadow Buttercup
(Ranunculaceae)

seemingly endless. For simplicity, a diagram of an imaginary, radially symmetric flower is used as an example. At its top the flower stem widens and, because it holds all the floral parts together, it is called the 'receptacle'. The 'calyx' – a circlet of petal-like structures, usually green or brownish, called 'sepals' – fits snugly under the 'corolla', the collective name for the petals. The corolla surrounds the 'androecium', which comprises a ring of 'stamens' (stamen = 'anther' plus 'filament'). The stamens encircle the 'pistil' (= 'carpel' or 'syncarpous gynoecium') comprising 'stigma' + 'style' + 'ovary'. If there are two or more pistils (carpels), they are termed a 'syncarpous gynoecium'. Inside the ovary are the 'ovules', only one ovule is depicted in the diagram. Each ovule encloses an 'embryo sac'.

Where is the pollen?

To see how pollen fits into the reproductive life cycle of the flower, it is worthwhile to look at the reproductive floral organs of a real flower. Florists' Lilies (*Lilium* cultivars) make good examples, and they are easily accessible to gardeners and non-gardeners. Lilies have hermaphrodite flowers, which is to say the male and female organs occur in the same flower – as they do in the diagram. The majority of plant species have hermaphrodite flowers. Lilies are easy to explore because the reproductive parts are quite large. By pulling a petal away the interior of the flower is revealed. In the centre is the pistil, the female organ of the flower, at the top of which is a tri-lobed stigma; below the stigma is the style, which is quite long. It has a central channel, the 'stylar canal' which leads from the stigma to the ovary 'locule', where the ovules are contained. There are six stamens surrounding the pistil; stamens are the male organs of flowers, and they are pivotal to the life cycle of pollen. In the Lily each stamen has a large anther held on a slender, stem-like filament. The most frequent numbers of stamens in flowers are three (or multiples of three), as in the lily), five (or multiples of five), or many. Inside the anthers are the pollen

grains. Most anthers have two sacs or 'thecae' (singular 'theca') separated by a 'connective' – the tissue connecting the two thecae of an anther. Each theca is separated into two compartments or 'locules'. At maturity the two thecae split open, usually longitudinally, to release the pollen grains from the locules in hundreds and thousands.

Stamens and pistils

The structure and form of stamens and pistils are remarkably varied. They are touched on only lightly here, because our focus is on pollen. Stamens have many different forms in flowering plants, apart from simple four-locular anthers on long, slender filaments. The filaments may be hairy or smooth, long or short, thick or thin; they may be basally united with short or long free filaments emerging from a coronet-like structure, or the filaments may even be completely united in a staminal tube. The basal fixing point may be on the petal – this is often the case in flowers that are in part tubular such as the mint family (Lamiaceae). In many families filaments are fixed basally on the receptacle. The tip of the filament may join onto the anther connective at the base, halfway up the connective, or at the top. The anther connective may be slightly visible, or highly elaborated and/or enlarged. Anther thecae are also very varied. Instead of the thecae opening via a longitudinal slit they may open horizontally, or even by pores. 'Poricidal anthers' are particularly associated with the Rhododendron family. Following opening ('dehiscence') the anthers may twist as in *Centaurium erythraea* (Centaury – Gentianaceae); or the thecae may be 'valvate' (have little hinged 'doors') as in the Witch Hazel – *Hamamelis*.

The stigma and style of the pistil also take many forms; for example in the Poppy the style is very truncated, and the stigma is like a miniature cartwheel. In *Mirabilis jalapa* (Marvel of Peru) as each flower opens the very long style is curled up like a 'catherine wheel', with a many-branched stigma at the tip. The styles of Tulip and Lily pistils are long and straight, with the stigma divided

Fuchsia cv. (Onagraceae) –
pollen grain, note viscin threads
[LM x 88, stained with Malachite green]

opposite: *Fuchsia* cv. (Onagraceae) –
close up of stamens and pistil

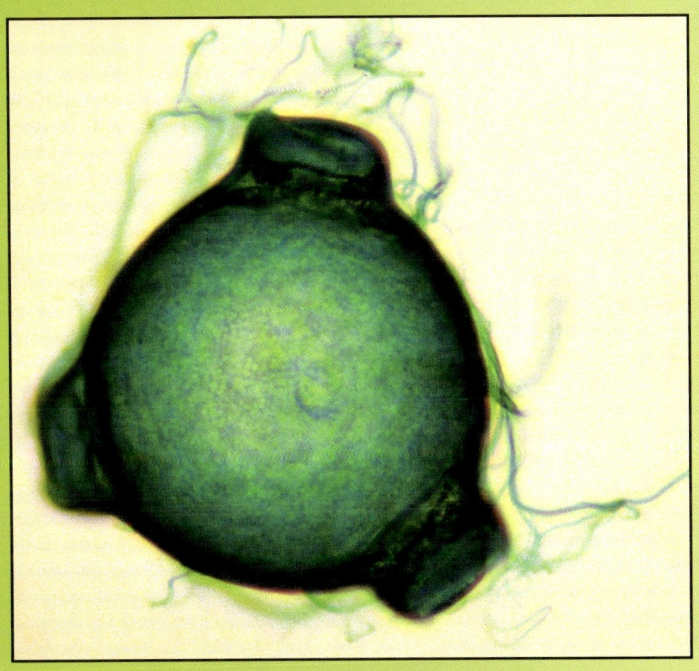

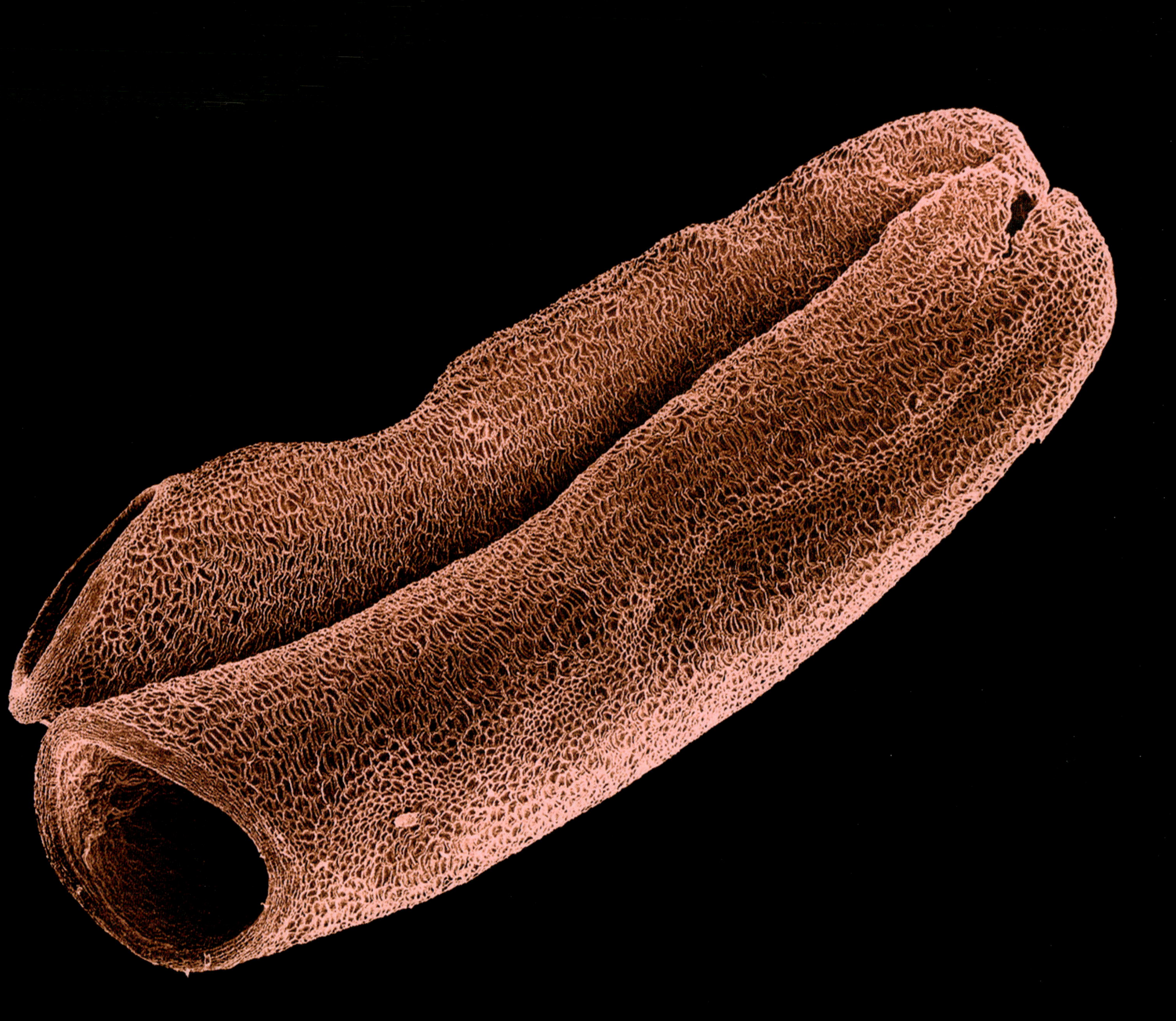

into three closely united segments. In Geraniaceae, five slender styles form a column from which the separate stigmas branch out in a delicate star formation, while in Malvaceae, for example *Hibiscus* and Common Mallow (*Malva sylvestris*), the style is surrounded by a staminal tube from which, at various points along the length of the tube, little anther-tipped filaments branch out. The branched stigma emerges from the top of the staminal tube, and the whole arrangement looks rather like a miniature Christmas tree.

THE DEVELOPMENT AND FUNCTION OF POLLEN

In most plants pollen grains are released from the anthers of mature flowers as individuals. However, in some plant families (circa fifty) there are at least some species where the mature pollen grains are dispersed as 'tetrads'. These include many members of the Heather and *Rhododendron* family (Ericaceae), as well as Evening Primroses, *Fuschia* and Rose-bay Willow Herb (Onagraceae). Pollen may also be shed as 'polyads'. In polyads the grains are normally in multiples of four. Polyads occur, for example, in *Acacia* and *Mimosa* (subfamily Mimosoideae, of the very large family Leguminosae).

Another pollen 'dispersal unit', the 'pollinium' occurs in, but is restricted to, two other very large families, the Orchidaceae and the Asclepiadaceae. Here the pollen grains are exposed in more or less compact coherent masses ('massulae'). There are usually four pollinia subdivided into four locules – these reflect the form of the more usual anthers in other families. The four pollinia may be together on a single branch ('caudicle'), or in pairs on two caudicles. The pollen masses are joined by the caudicle to an adhesive structure called the 'viscidium'; the complete structure is referred to as a 'pollinarium'.

Pollen grains develop within the developing anthers from specialised 'sporogenous cells', which are surrounded by a protective inner layer of the anther wall – the 'tapetum'. The 'pollen mother cells' develop from 'diploid' sporogenous cells; subsequently each diploid pollen mother cell undergoes

Rhododendron cv. – stamens with poricidal anthers

opposite: *Rhododendron* cv. – a poricidal anther, the pores are at the top of the thecae [SEM x 30]

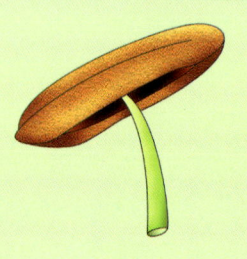 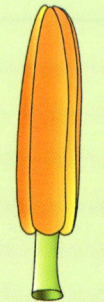   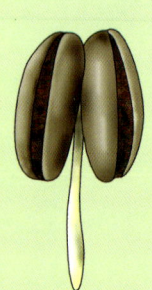

Examples of differing stamen forms and structures
[redrawn from Kerner and Oliver (1903) and Lawrence (1955)]

'meiosis', dividing into 'haploid' 'daughter cells', usually four – the tetrad – or, less commonly, a polyad or a pollinium. Following meiosis, the individual pollen cells within the tetrad, polyad or pollinium, continue to develop. In most flowering plants, prior to maturity, tetrads separate into individual pollen grains, while polyads or pollinia remain intact. Inside each young fully formed pollen grain, before its release from the anther, the pollen nucleus undergoes 'mitosis', dividing to provide the mature pollen grain with two cells: the 'vegetative cell' containing the 'vegetative nucleus', and the 'generative cell' containing the 'generative, or sperm, nucleus'. Unless stored under special conditions, most pollen grains are generally very short-lived, a few days at best. Therefore, following release from the anther the pollen grains must be dispersed rapidly, while they are still viable, to fertilise other flowers of the same species.

Pollen grains are not parts of a plant like petals, leaves, stamens or stems; they are individuals – neat, highly functional self-contained structures; the haploid generation of the flowering plant life cycle, the microscopic counterparts of the more obvious diploid generation – the whole plant. The haploid (a cell which participates in sexual fusion) gametophytic generation contains only one representative of each chromosome, of the chromosome complement, while the diploid generation has double the haploid number of chromosomes in the nuclei of its non-reproductive ('somatic') cells. When a pollen grain leaves the anther of a diploid parent plant, and arrives on the

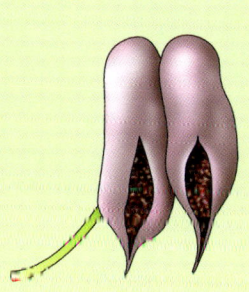
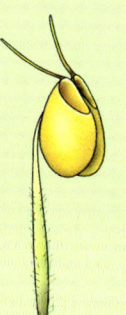

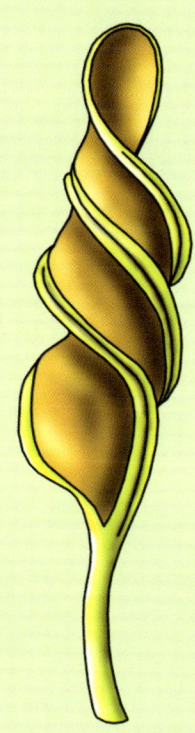

stigma of another diploid plant of the same species, it will germinate and the sperm cells will fuse with the egg cell nucleus and polar nuclei of the ovule (this will be described in more detail later). The fertilised ovule will subsequently mature into a diploid seed, which will then develop into a diploid plant.

The difference between pollen and spores

There is a fundamental difference between pollen grains and spores. Alternation of generations (haploid and diploid) is unique to plants, from green algae, through liverworts, mosses, ferns, conifers and their relatives (gymnosperms) to flowering plants (angiosperms). It has no counterpart in the animal kingdom.

However, with the exception of the conifers and their relatives, which technically speaking have pollen grains, non-flowering plants produce spores. Unlike flowering plants and gymnosperms, spore-producing plants also have an asexual generation, the 'sporophyte'. For example, on the underside of a mature fern frond – such as the male fern *Dryopteris filix-mas* – the individual leaflets (pinnae) have rows of tiny kidney-shaped structures, each one of which is a 'sorus' (plural 'sori'). Protected under each sorus are the sporangia, containing the spores. When the sporangia are mature they burst open to release the spores which, like pollen grains, are the haploid generation. After dispersal from the sporangia of the diploid parent plant the spores germinate,

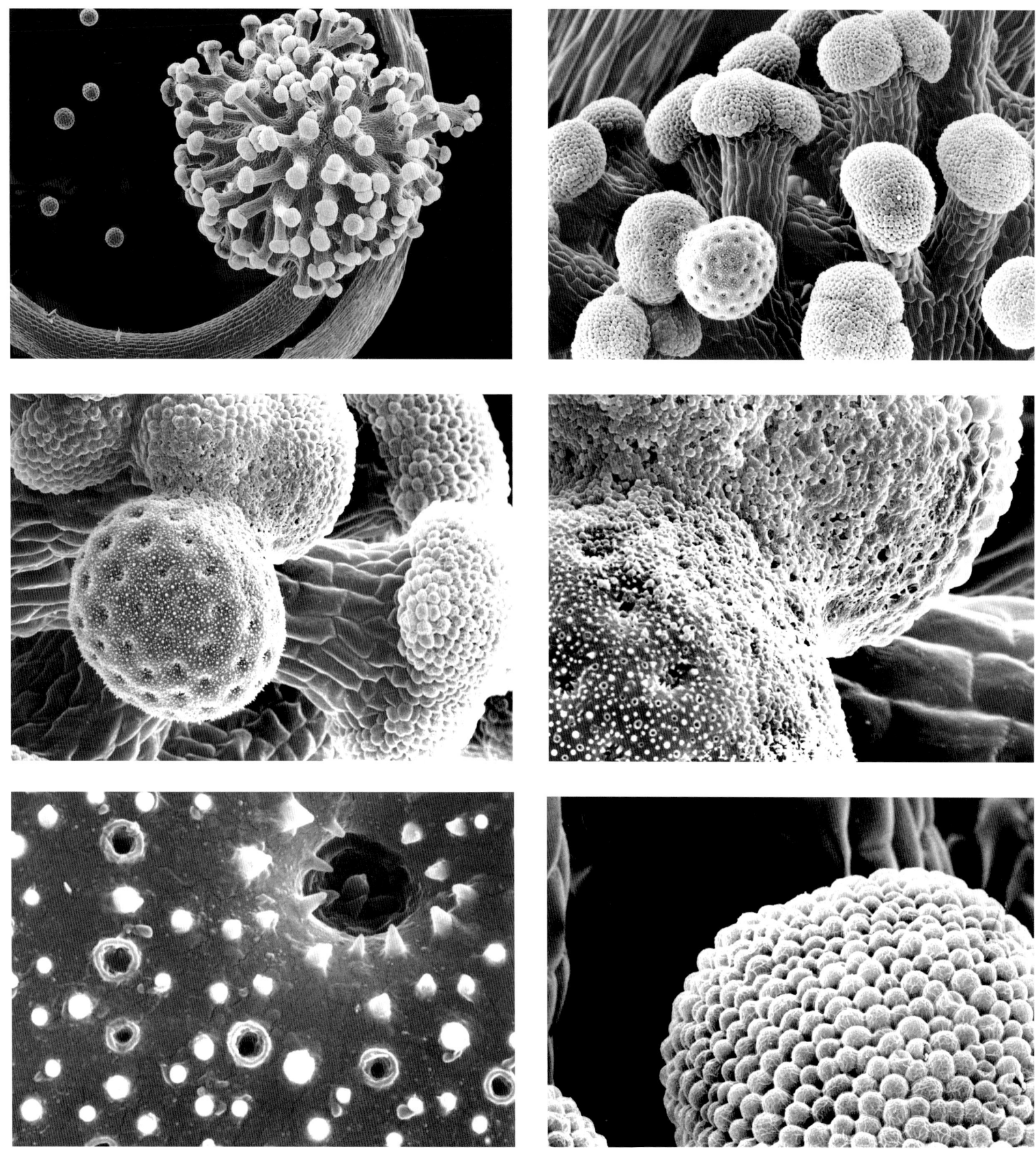

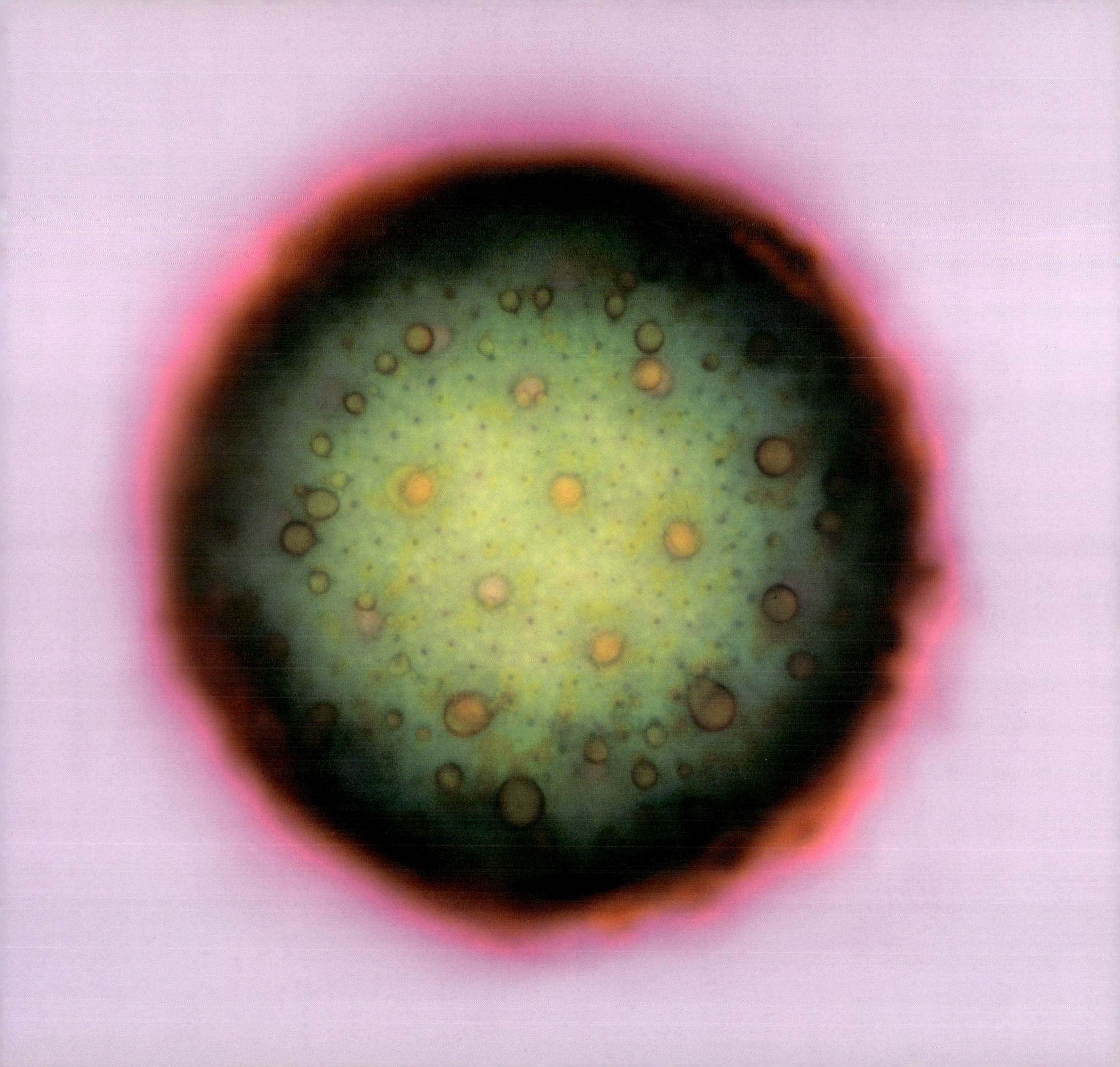

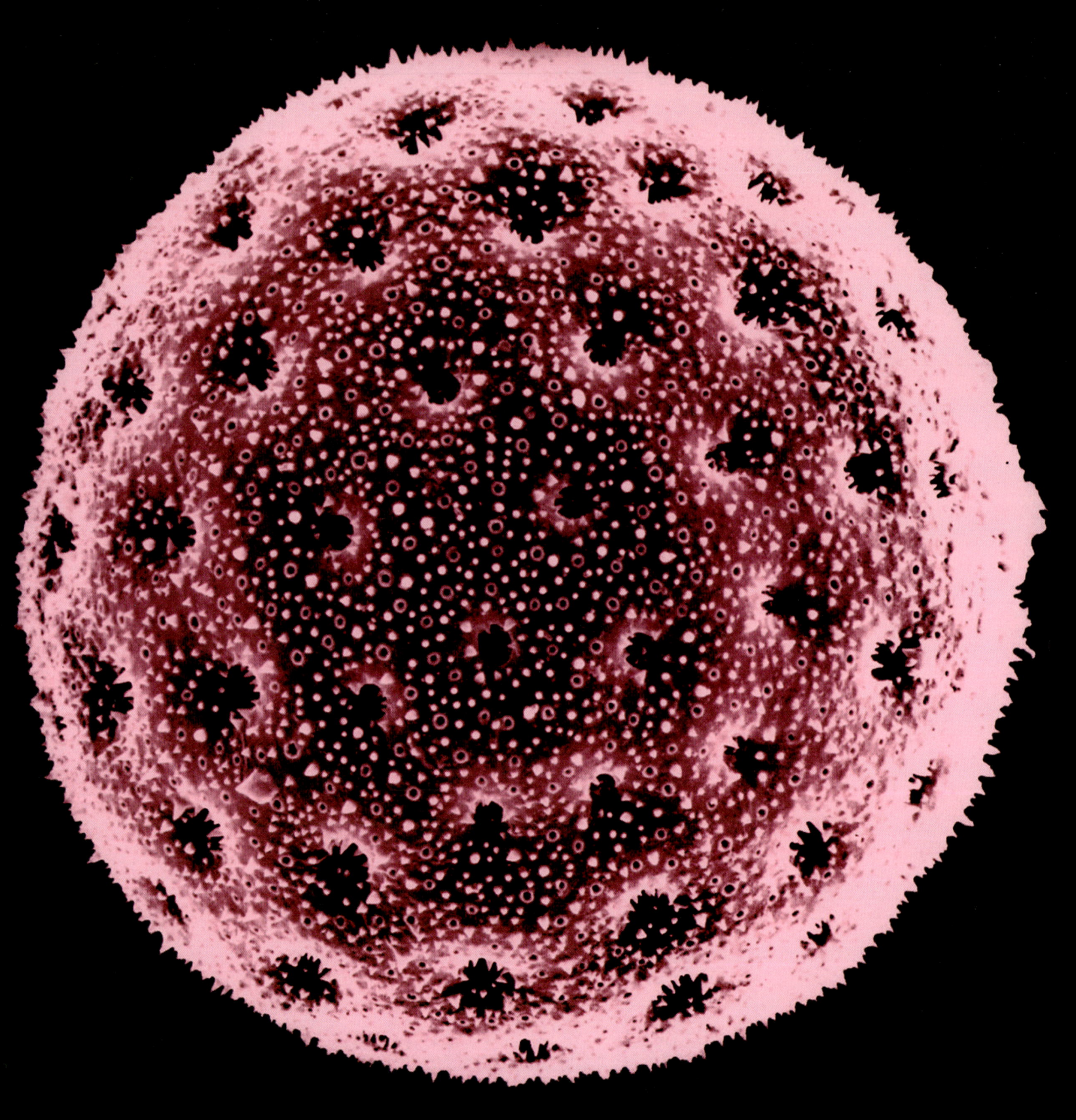

in moist conditions, and develop into the tiny organ of the haploid generation, the 'prothallus'. On its underside, the prothallus has a female egg-bearing organ – the 'archegonium' – and a male sperm-bearing organ – the 'antheridium'. The sperm, dependent on a very damp environment, leave the antheridium of one prothallus to swim across to the archegonium of another prothallus to fertilise an egg cell, and also ensure cross-fertilisation. The result of this union is a diploid sporophyte, which after a few years will grow into a mature fern.

POLLEN STRUCTURE

To the human eye pollen grains are fascinating and beautiful structures but, like any natural phenomenon, our concept of beauty is based on aesthetic perception not on function and efficiency. However, pollen grains are not only functional but also very efficient.

Much of what we find beautiful in plants and animals is, nevertheless, highly relevant to the evolutionary scheme. This is because our senses select natural 'advertisements', evolved to attract one animal to another for mating or food, an insect to a plant, or a bird to an insect, and so forth.

The human race, with its finely developed sensibilities, is a very late developer in the evolutionary clock. Insects and other invertebrates, fish, snakes and lizards, birds, fungi, mosses, ferns, flowers and other plant groups, as well as other non-mammalian groups, the mammals, and the natural landscape in which they live, had all evolved long before us.

This succession of evolution is reflected in the Bible, where we read in Genesis 1–3 the beautiful allegorical story, *In the Beginning.* Firstly God put in place the Earth to divide the Seas, then the plants, and then all the birds, sea-dwelling creatures, and animals. Lastly he made Adam and Eve "in his own image", and planted for them the Garden of Eden, where they were to encounter the 'Tree of Knowledge of good and evil', and a serpent to tempt them to taste the ripe fruits of the Tree.

previous page left: *Mirabilis Jalapa* – Marvel of Peru (Nyctaginaceae) – top left: multiple-headed stigma, note coiled style below [SEM x 30]; top right: close up of stigma with a pollen grain attached to one of the stigma heads [SEM x 200]; centre left: close up of pollen adhering to stigma head, note similarity of size between pollen grain and mini stigma head [SEM x 400]; centre right: even higher magnification to show pollenkitt adhesion between pollen and stigma [SEM x 1000]; bottom left: surface of pollen grain with one small circular aperture showing [SEM x 6000]; bottom right: surface of stigma [SEM x 1000]

previous page right: *Mirabilis Jalapa* – Marvel of Peru (Nyctaginaceae) – fresh pollen grain [LM x 40, stained with a mixture of Malachite green, Acid Fuchsin and Orange G]

opposite: *Mirabilis Jalapa* – Marvel of Peru (Nyctaginaceae) – pollen grain [CPD/SEM x 500]

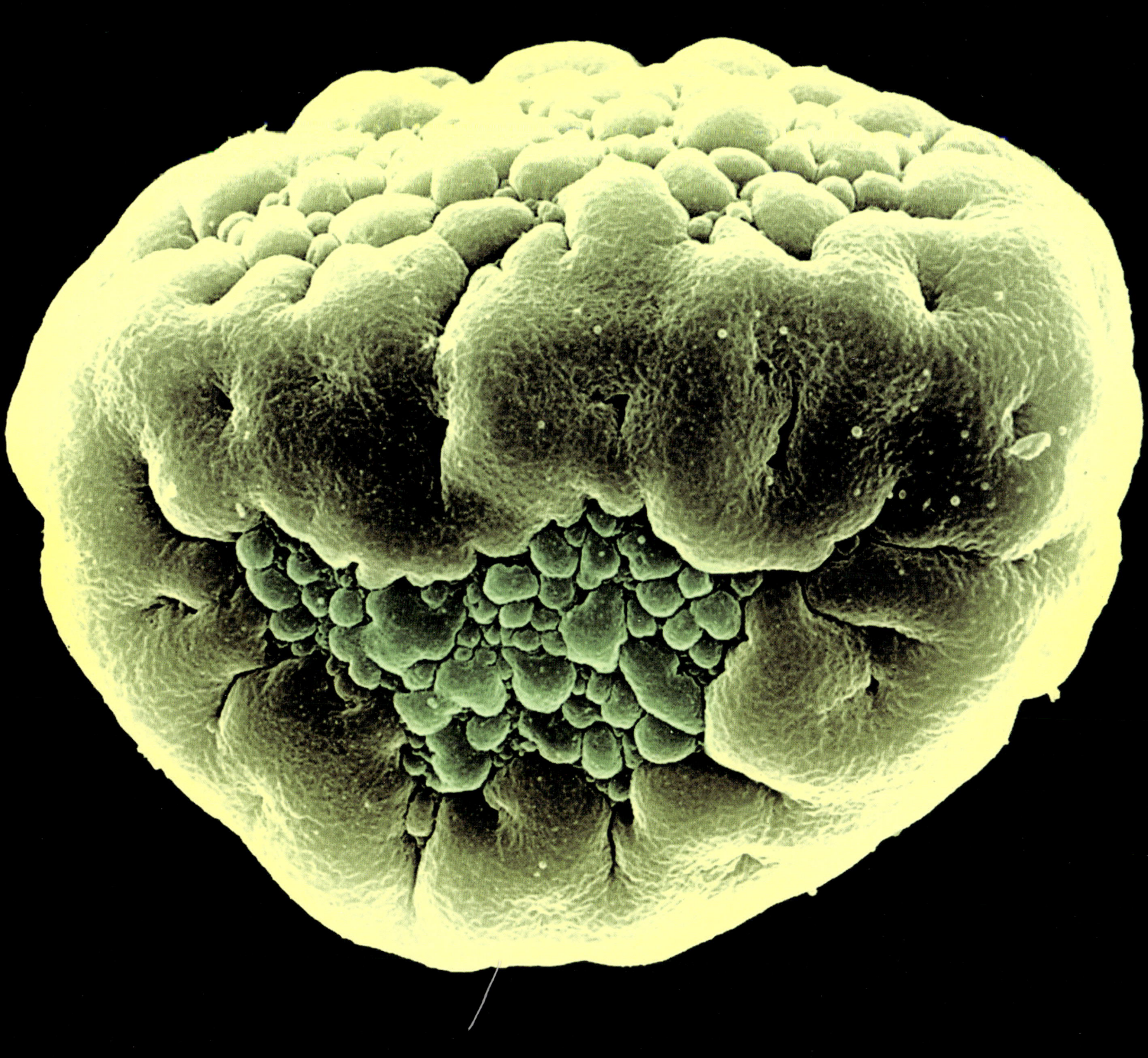

The pollen wall

We use the words 'architecture' and 'ornamentation' in the description of pollen grains only because they help us to describe what we see or envisage. The outer pollen grain wall, the 'exine', which in many species is very elaborate, is composed of a substance called 'sporopollenin'. The often elaborate 'exospore' of the spore walls in non-flowering plants such as ferns, mosses and liverworts also consists of sporopollenin. Sporopollenin is one of the toughest plant substances known, the diamond of the plant world.

Such a substance needs at least a brief explanation. The precise chemical nature of sporopollenin has never been unequivocally characterised. Basically, it consists of carbon, hydrogen and oxygen in an approximate ratio of 4:6:1; fatty, aromatic and minimal carboxylic acids are also present. Apparently, although these components are consistent through all plant groups, their ratios differ. It has been suggested that sporopollenin is probably a randomly cross-linked biomacromolecule without a repetitive large-scale structure, and that this characteristic would make it inherently resistant to enzymic attack, as well as to many laboratory procedures designed to reduce/return it to its principal components. This would account for the extraordinary preservation qualities of the pollen exine. Given the right conditions for burial and preservation, this tough outer casing can resist decay and remain structurally unaltered in a 'mummified state', for millions of years, thus providing a rich pollen and spore fossil record from which we have been able to deduce much of our current knowledge on the evolution of non-flowering and, later, flowering plants.

Beneath the sporopolleninous exine is an inner layer, the 'intine'. The intine surrounds the 'cytoplasm', which fills the interior of the fully formed pollen grain prior to dispersal. The vegetative and generative cells are enclosed within the cytoplasm, together with all the other specialised cellular components ('organelles') responsible for pollen function. To make it easier to visualise in three dimensions imagine a more or less spheroidal iced fruit cake where the

page 38: *Pelargonium zonale* cv. - Geranium (Geraniaceae) – centre of flower showing star-like, five-branched stigma.

page 39: *Papaver orientale* – Oriental poppy cultivar (Papaveraceae) – centre of a radially symmetrical flower; note cartwheel-like stigma surrounded by numerous stamens.

page 40: *Hibiscus rosa-sinensis* – Rose of China (Malvaceae) – a staminal tube with little stamens branching off; the style is enclosed by the tube but the branched stigma can be seen at the top of the tube.

page 41: *Abutilon pictum* (Malvaceae) – opened anther curling back to release pollen grains.

previous page left: *Tulipa* cv. – Florist's Tulip (Liliaceae) – showing the six anthers typical of Tulips, and many other monocotyledons, and the pistil with a trilobed stigma.

previous page right: *Tulipa armena* (Lilaceae) – pollen grain [CPD/SEM x 500]

opposite: *Tulipa violacea* – Tulip (Liliaceae) – pollen grain, expanded condition [CPD/SEM x 1500]

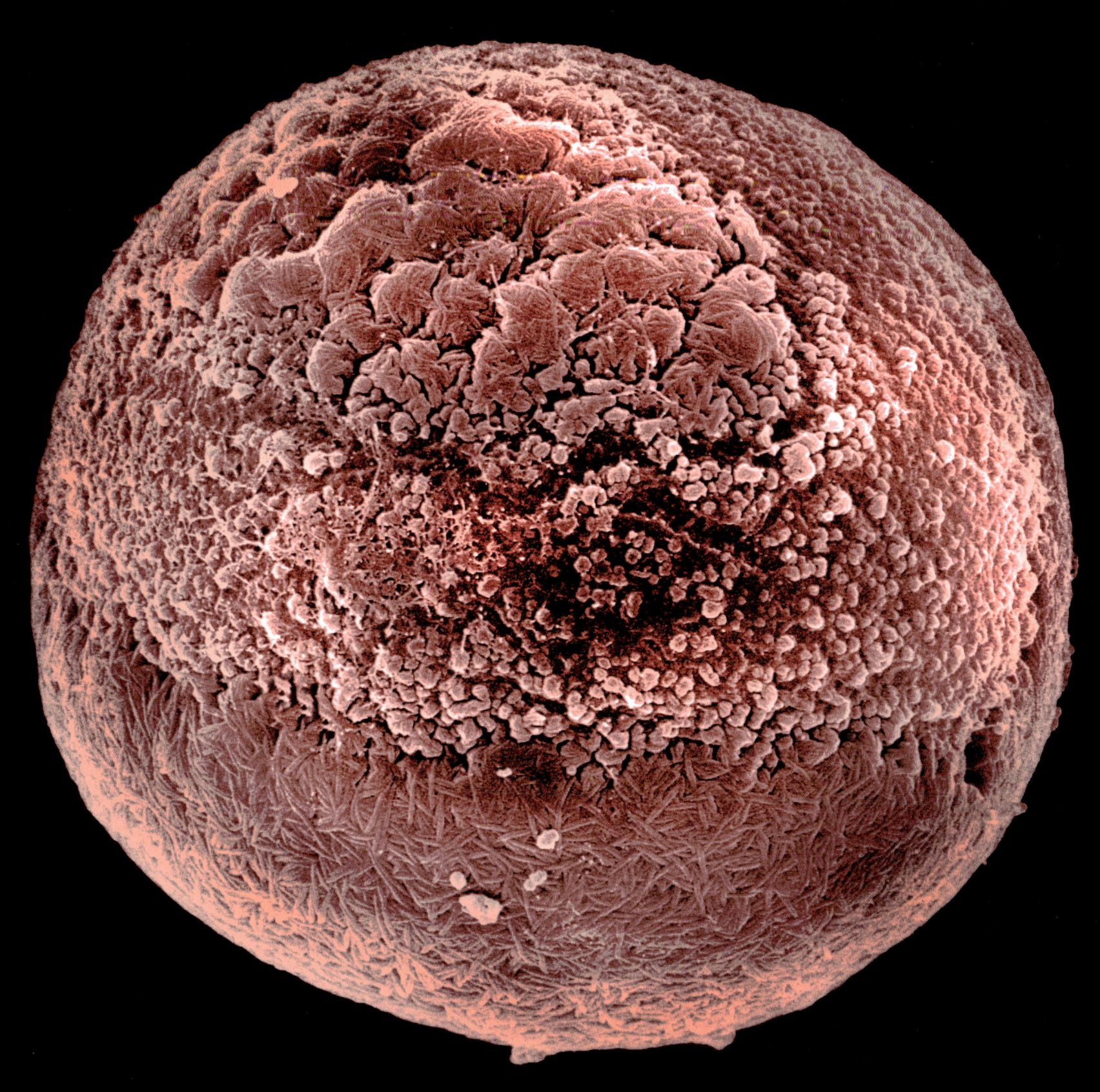

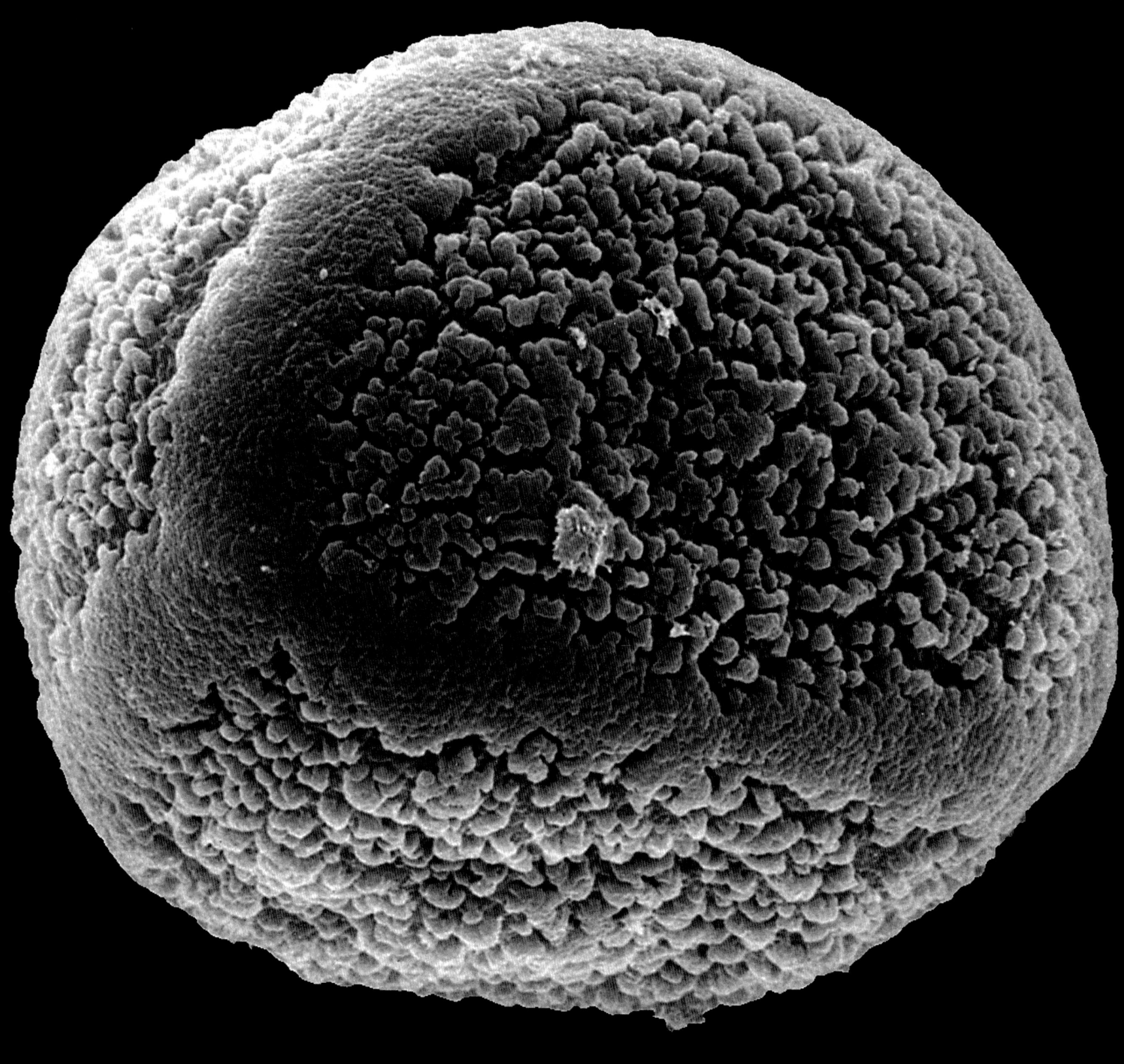

46

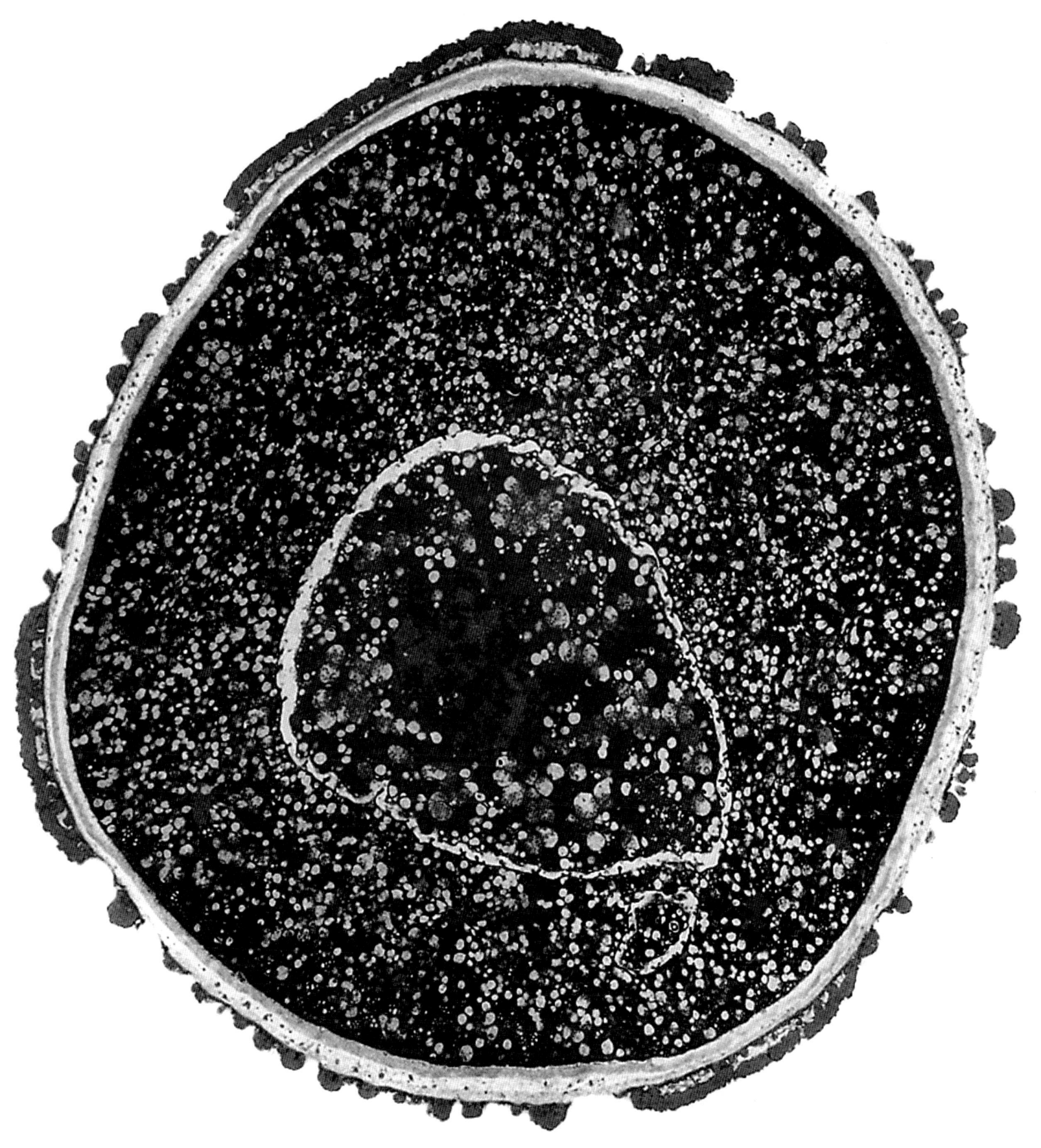

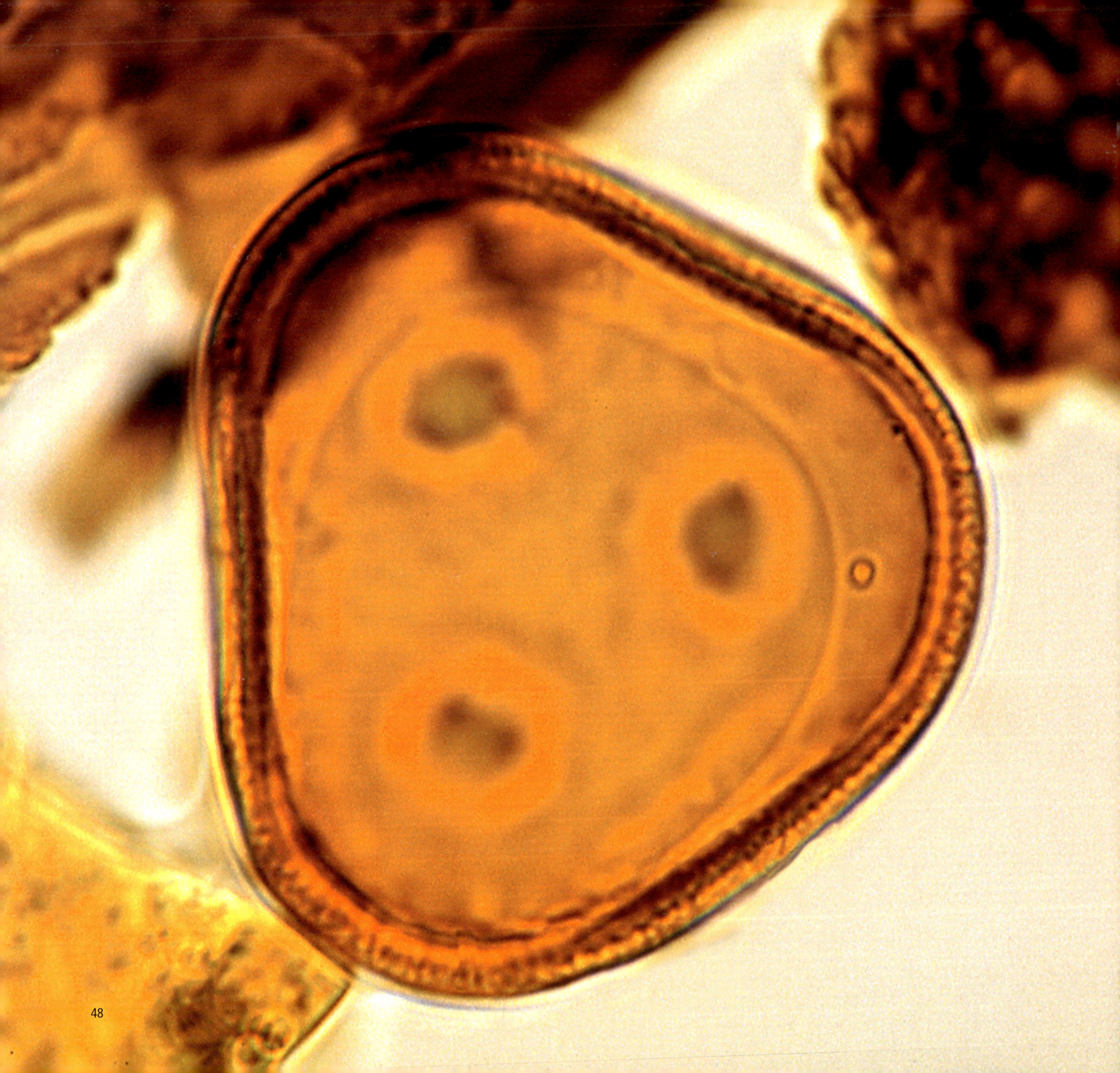

exine is the white icing, the intine is the marzipan and the cytoplasm is the rich cake mix in which the cellular organelles are distributed (represented in the imaginary fruit cake by dried fruits, cherries and nuts). The intine and the exine both have a protective role, but only the exine is tough and corrosion resistant because it has to protect the vegetative and generative cells during transit to the stigmatic surface of the pistil. Later the intine protects the sperm cells during pollen germination.

Characteristically, the outer wall of the pollen exine has two major layers, the 'endexine' and the 'ectexine'. The endexine is the innermost of the two layers, and although it is corrosion resistant its composition, which is fairly uniform, differs somewhat from ectexine. The outer ectexine is typically subdivided into three zones: directly above the endexine layer there is the 'foot layer', above the foot layer the 'columellar layer' (or 'infratectum'), and above this the 'tectum' (from the Latin for roof). The columellar layer is so named because it is often composed of many columns ('columellae') as in a Greek temple. There are numerous pollen wall types and, of these, many in which either the tectum or the foot layer is absent. The ornamental elements on the pollen exine may be on top of the tectum ('supratectal'); where the tectum is absent, the ornamental elements are modified columellae. The extraordinary ranges of pattern and of ornament that we see, when we look at the surface of pollen grains under the microscope, are all modifications of the tectum or of the exposed columellae.

Pollen apertures

The other very important characteristic of most pollen grain exines is the presence of one or more openings for the exit of the germinating pollen tube, which carries the reproductive cells from the pollen grain to the ovule. These openings, or 'apertures' are covered by a thin membrane – the 'aperture membrane' – designed to burst under pressure (cf. the eardrum). This

previous page left: *Tulipa kaufmanniana* – Tulip (Liliaceae) – a pollen grain with three large apertures [CPD/SEM x 2000]

previous page right: *Tulipa vvedenskyi* – Tulip (Liliaceae) – an ultrathin section through a pollen grain showing, from the outside, the exine (the dark grey layer), the intine (pale grey layer), and the cytoplasm with organelles (speckled inner area), with well-developed generative cell in central area. The three thicker areas of the exine correspond to the banded regions seen in the previous image, while the thinner areas of exine correspond to the crumbly surface of the large apertures in the previous image [TEM x 1000]

opposite: *Trilatiporites* – a triporate fossil pollen grain of unknown affinity focussed to show exine surface detail. The three pores are visible, but not in sharp focus. From the Neyveli lignite of India, c. 22 million years old (early Miocene). Other microscopic pieces of fossilised plant 'debris' are also apparent [LM x 1000 – natural fossilised colour – cf . with acetolysed pollen grain colour]

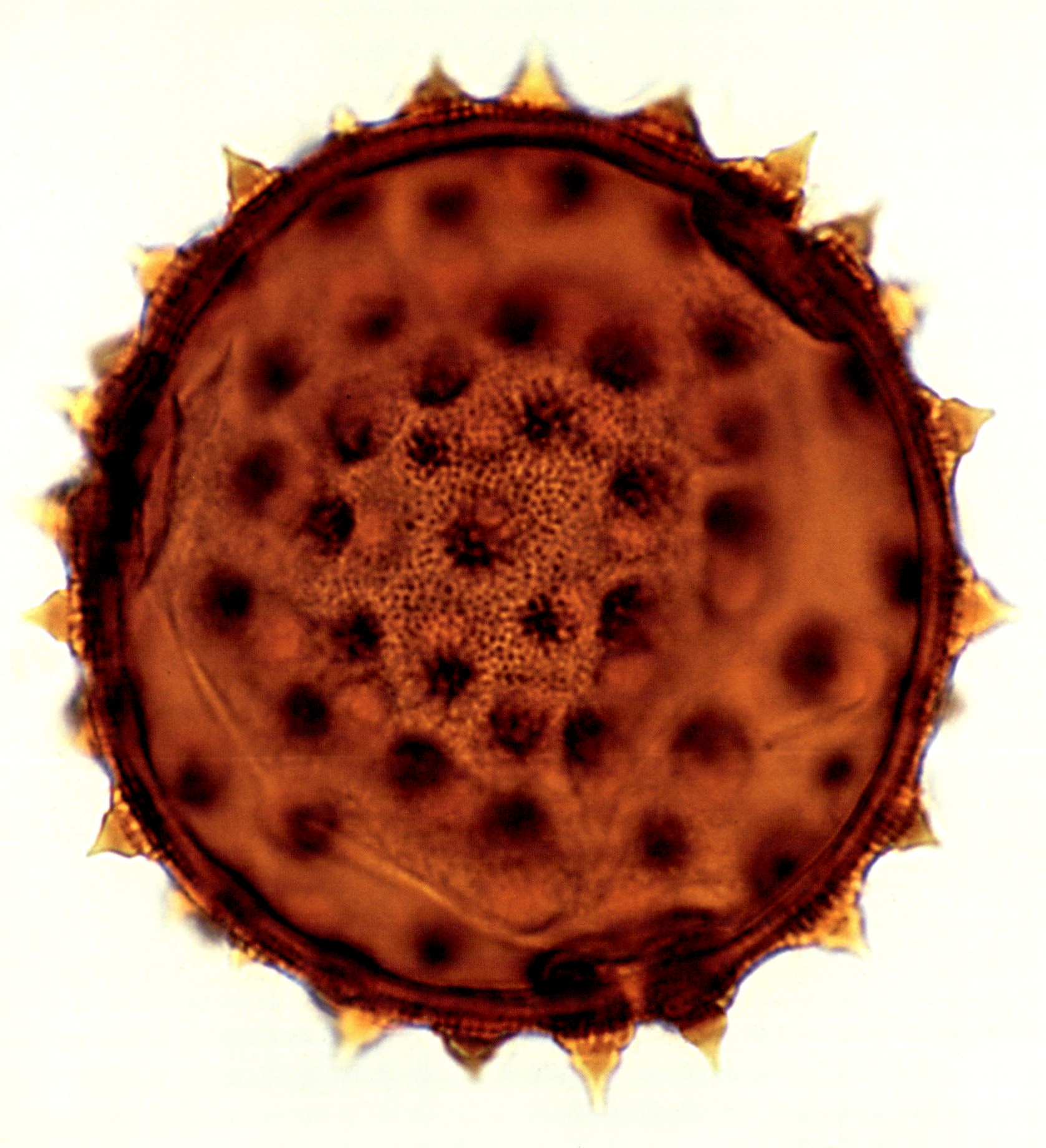

50

membrane is usually destroyed by acid treatment, and is almost always absent in fossil pollen grains. The number of apertures in a pollen grain varies, among different species of plant, from one to many. The earliest fossil pollen grains recovered, from about 120 to 130 million years ago, have just one elongated slit-like aperture, and many flowering plants such as the magnolias and the palms, still carry this characteristic. Both groups of plants represent very early evolving flowering plant families. However, pollen grains with three simple, radially distributed elongate apertures are also found very early in the pollen fossil record, and this aperture arrangement is still found in many living plants such as the Christmas Rose (*Helleborus niger*), Witch Hazel (*Hamamelis*) and *Acer* (Maples, Sycamore).

Why does pollen have so many variations?

Pollen grains have a clearly defined role in the life cycle of the plant. Why such an array of fantastic modifications to the original simple mono- and tri-aperturate 'prototypes' should have developed is not easy to explain other than as evolutionary adaptations or developmental modifications. One argument, often put forward, is that a greater number of apertures provides a wider choice of exit for the developing pollen tube thus giving it a greater chance of leaving via an aperture in contact with the stigmatic surface, with a resultant saving in energy and materials for the rapid growth phase to follow. However, pollen with the least specialised aperture formations – one or three slit-like apertures – from which all other aperture types and arrangements have evolved, still occur in many flowering plants, and they germinate successfully. Probably the most significant example of successful mono-aperturate pollen is in the grasses. They are a late-evolving but highly successful and rapidly diversifying group of plants of great economic importance, and dominate a wide variety of habitats throughout the world. All species of grass, of which there are thousands, have pollen with just one small pore. It is possible that

Abutilon cv. – 'Cynthia Pike' (Malvaceae) – pollen grain, focussed to show pollen wall detail. Note the three aperture sites in the wall. [LM x 400 – acetolysed]

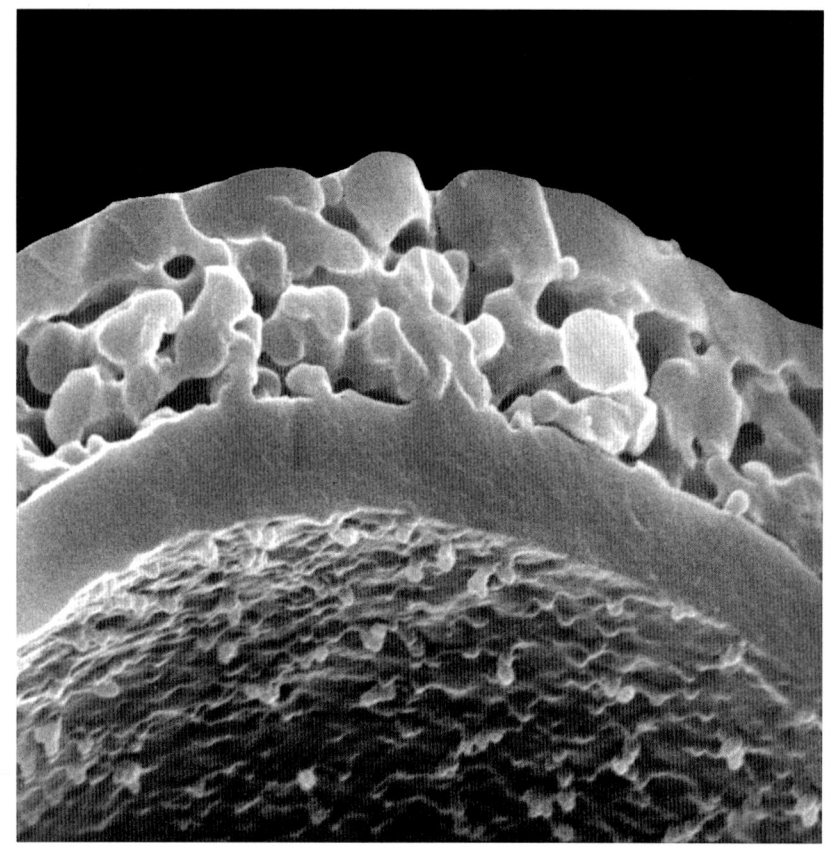
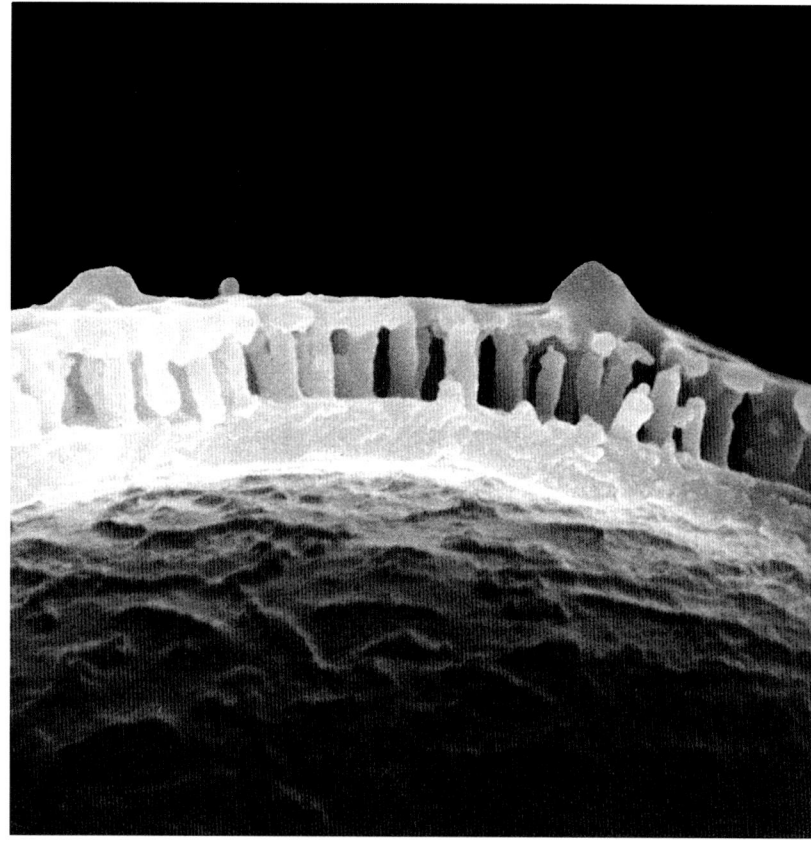
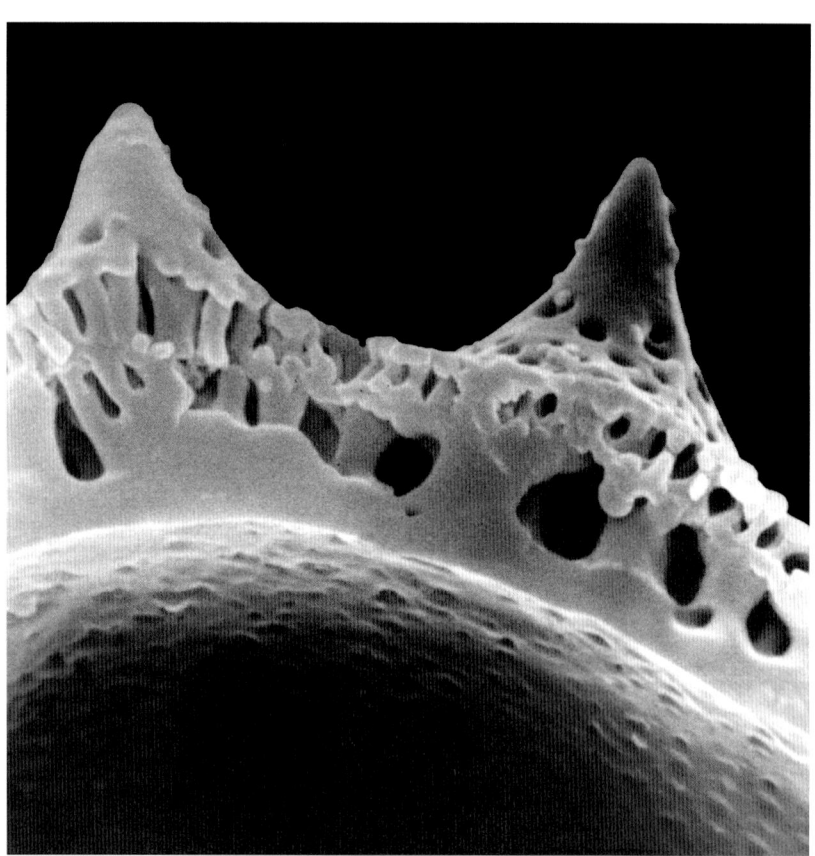

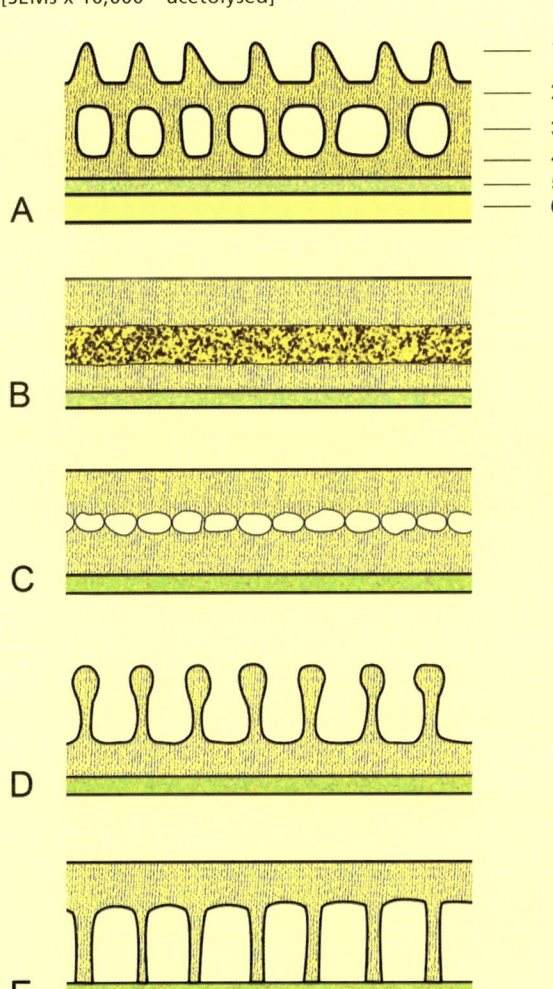

pollen wall structure: A) typical pollen wall and basic wall layer terminology – 1. supratectal structure, 2. tectum, 3. infratectum, 4. foot layer, 5. endexine, 6. intine; B-E) some variations in pollen wall structure: B) infratectum spongy rather than columellate; C) infratectum narrow, tectum and foot layer comparatively wide; D) tectum reduced to slightly expanded heads on the columellae; E) well-developed tectum, long columellae, no foot layer – bases of columellae sit directly on endexine. NB. Intine is not represented in B-E because the diagrams are based on acetolysed pollen, which destroys intine.

opposite: Pollen fractures to show some of the variation in wall structure, and tectum, infratectum and foot layer thickness:
top left: *Morina longifolia* (Dipsaceae); top right: *Adansonia digitata* – Baobab (Bombacaceae); bottom left: *Achillea millefolium* – Yarrow or Milfoil (Compositae); bottom right: *Malva sylvestris* – Common Mallow (Malvaceae) [SEMs x 10,000 – acetolysed]

monoaperturate pollen grains may even be weighted to stand a higher chance of falling aperture side down onto the receptive stigmatic surface, although pollen grains with three slit-like apertures, each with a central pore, are the commonest of all pollen types. More explicable are pollen adaptations to a particular type of environment inhabited by the parent plant; for example, the filamentous pollen of water-dwelling plants such as Eel Grasses (Zosteraceae), or plants that distribute their pollen on the wind or across water, like the conifers, which have inflated 'buoyancy bags' on their pollen grains. Nevertheless, as new plants have evolved so new pollen types have also evolved, often highly specific to the parent plant.

Pollen size

Pollen grains are microscopic; their detail cannot be resolved by the naked human eye unless they are at the larger end of the size range. They are measured in microns – a micron is one thousandth of a millimetre. Most pollen grains are between 20 and 80 microns. However, the smallest pollen grains are about 5 to 8 microns, as in some species of the Forget-me-not family (Boraginaceae); the largest recorded are more than 500 microns, although these are very unusual. More accessibly, some species in the cucumber and squash family (Cucurbitaceae) may have pollen grains with a diameter of as much as 250 microns.

Pollenkitt

In their natural state pollen grains may be dry or sticky. Dry or powdery pollen grains are frequently associated with wind-pollinated plants such as Birch, Alder, Hazel, Oak, Stinging Nettles and grasses, some of the plants whose pollen may make us sneeze in spring and early summer. Sticky pollen grains are associated with pollination by insects, birds, or other animals. The stickiness comes from a coating of oily lipids ('pollenkitt'). Pollenkitt has many

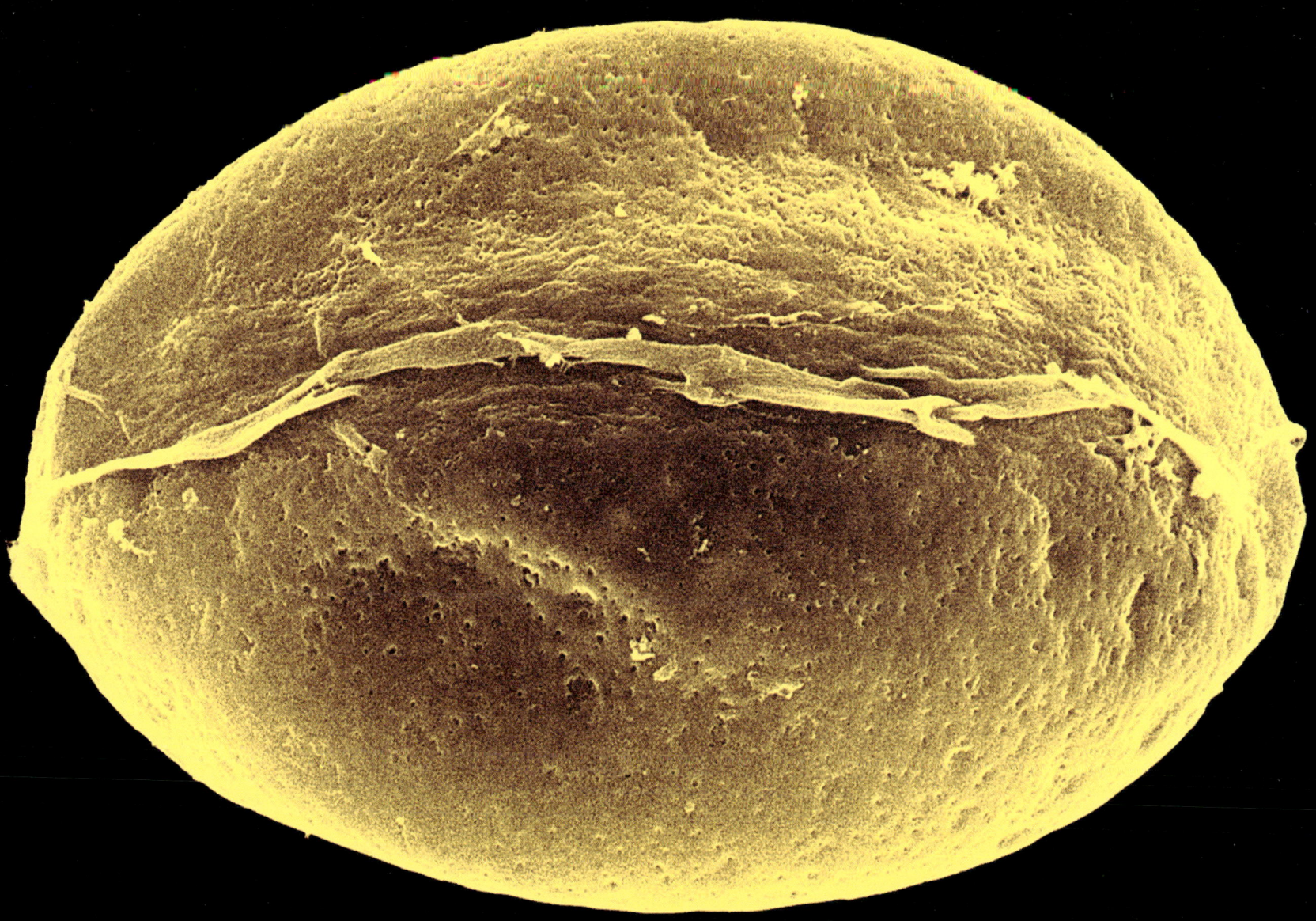

Magnolia soulangiana (Magnoliaceae) – a pollen grain,
expanded condition; a very early evolving pollen form
found near the base of the flowering plant fossil record.
The single aperture is almost imperceptible; the fine ridge
along the centre top of the grain is the slightly wrinkled
central area of the very thin aperture membrane.
[CPD/SEM x 2000]

opposite: *Magnolia cylindrica* (Magnoliaceae) – showing the
reproductive parts (central gynoecium surrounded by
numerous stamens) of the hermaphrodite (bisexual) flower.

overleaf left: *Hamamelis mollis* – Witch Hazel
(Hamamelidaceae) – cluster of flowers, with floral parts in
fours; note the dark reddish sepals, long narrow yellow
petals, and four central anthers.

overleaf right: *Hamamelis mollis* – Witch Hazel
(Hamamelidaceae) – pollen grains in different views; top
and bottom left: equatorial views; bottom right: polar view.
[CPD/SEM x 3000]

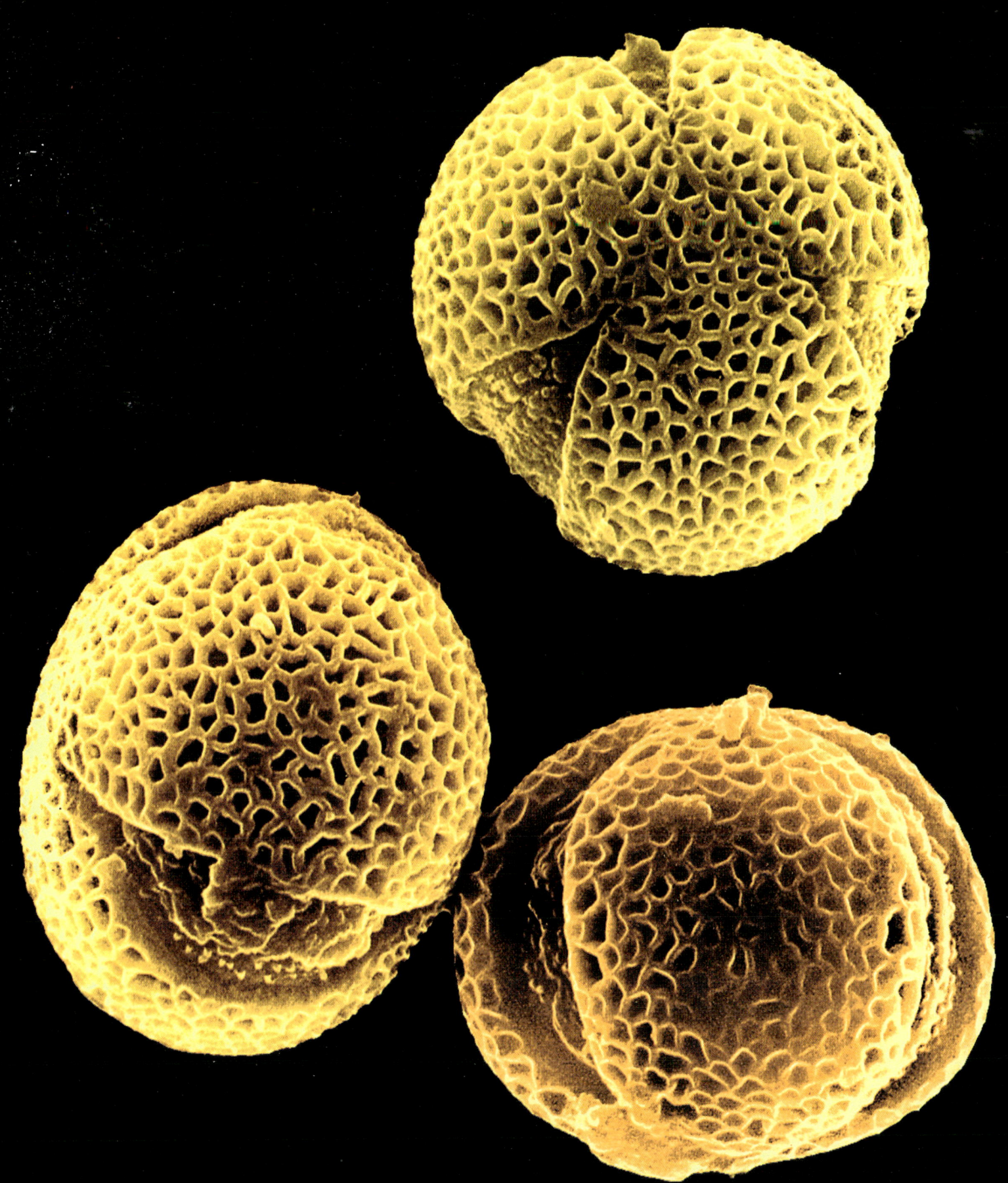

Hamamelis mollis – Witch Hazel (Hamamelidaceae) – closed
anther; the locules have door-like openings to release
pollen. [CPD/SEM x 40]

opposite: *Hamamelis mollis* – Witch Hazel (Hamamelidaceae)
– open anther; note the pollen inside the locule.
[CPD/SEM x 40]

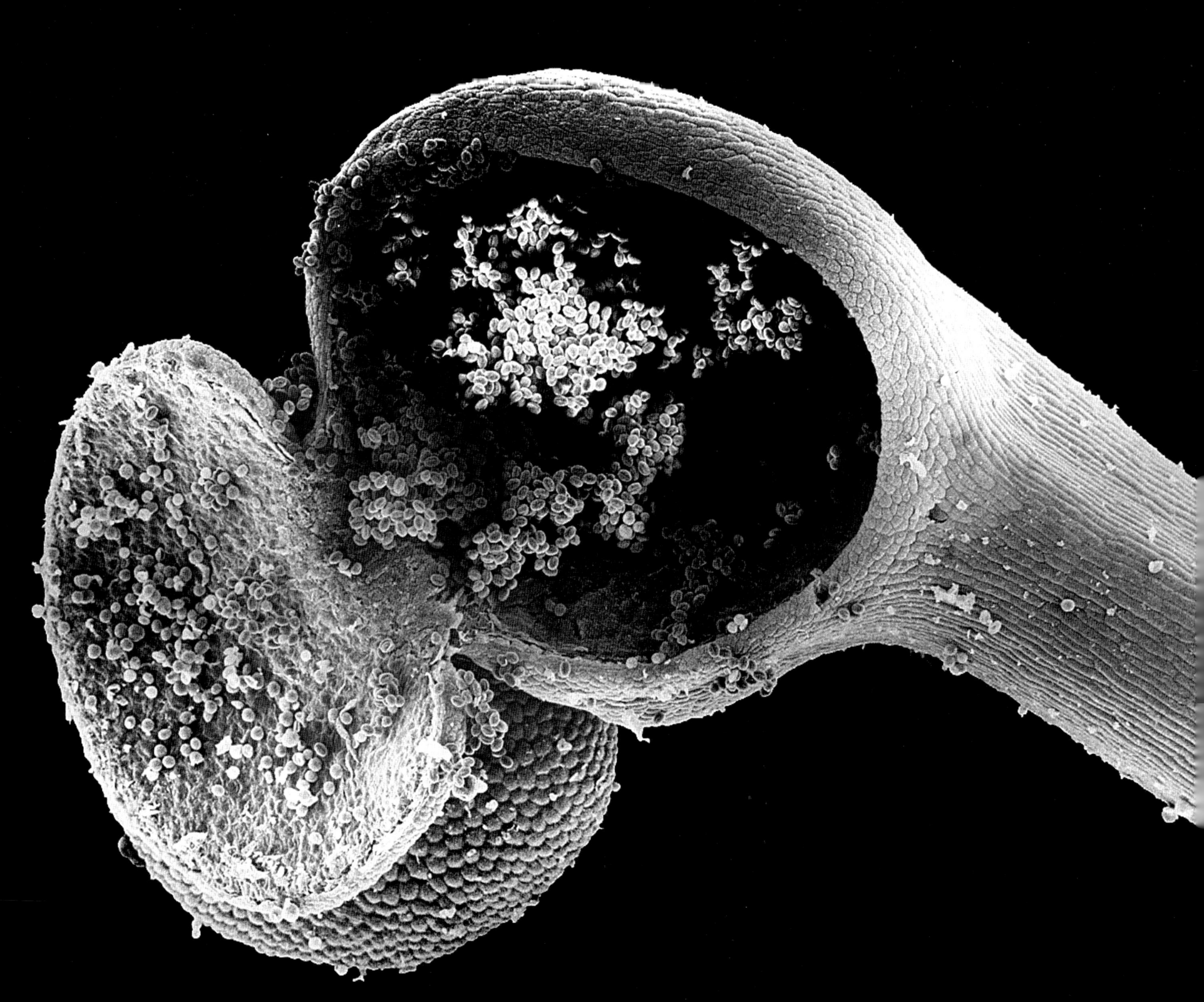

functions, including protection of the cytoplasm from solar radiation, maintaining the pollen proteins inside the exine cavities, attracting insects, and allowing pollen to adhere to body parts of pollinating visitors.

Pollenkitt is produced by the protective inner layer of the anther wall (tapetum) and deposited on the pollen grains as the anther and pollen reach maturity. Some dry pollens associated with wind pollination, such as in the grasses, have now been demonstrated to have a thin deposit of pollenkitt, which supports some of the range of functions, other than insect attraction, and attachment, which have been suggested for this substance. Easily available flowers in which to observe pollenkitt, include Florists' Lilies (*Lilium* cultivars) and Day Lilies (*Hemerocallis* cultivars), where the pollen grains are so well-coated by yellow or orange oily lipid that if brushed against they stain our skin or clothing. Indeed many retailers of these flowers attach warning labels to their pre-packed bouquets. Some even go as far as picking the pollen-rich anthers off the filaments in Lilies to avoid customer complaints!

Pollen: natural colours

Although most frequently ranging from colourless to yellow, fresh pollen grains – depending on the species of plant – may have one of a variety of colours. The pigments are mainly carotenoids or flavonoids. Carotenoids range from light to deep yellow or orange, while the flavonoids include colourless to yellow flavones and isoflavones, as well as red and purple anthocyanins. Either type of pigment may be contained in the sporopolleninous exine, but flavonoids seem to predominate, while carotenoid pigmentation predominates in the external oily lipidic coating as, for example, in Lilies. The amount of flavones and anthocyanins, and their location, all contribute to the visible colour spectrum of pollen. For example, identical flavonoids may be present in pollen of two species but because of variation in concentration may appear different in colour. Complexing of flavonoids with aluminium or iron,

Helleborus orientalis – Hellebore (Ranunculaceae) –
pollen grain in polar view. [CPD/SEM x 2000]

opposite: *Helleborus orientalis* – Hellebore (Ranunculaceae)
– a dark form of this very variably coloured garden cultivar

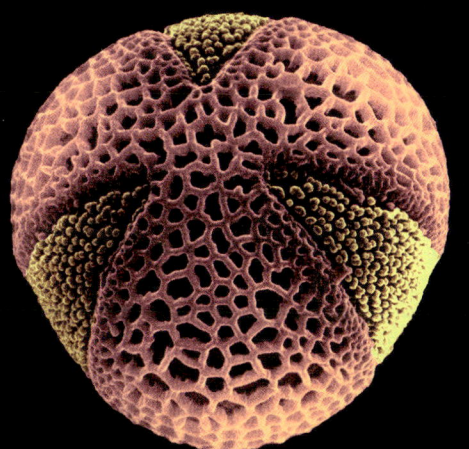

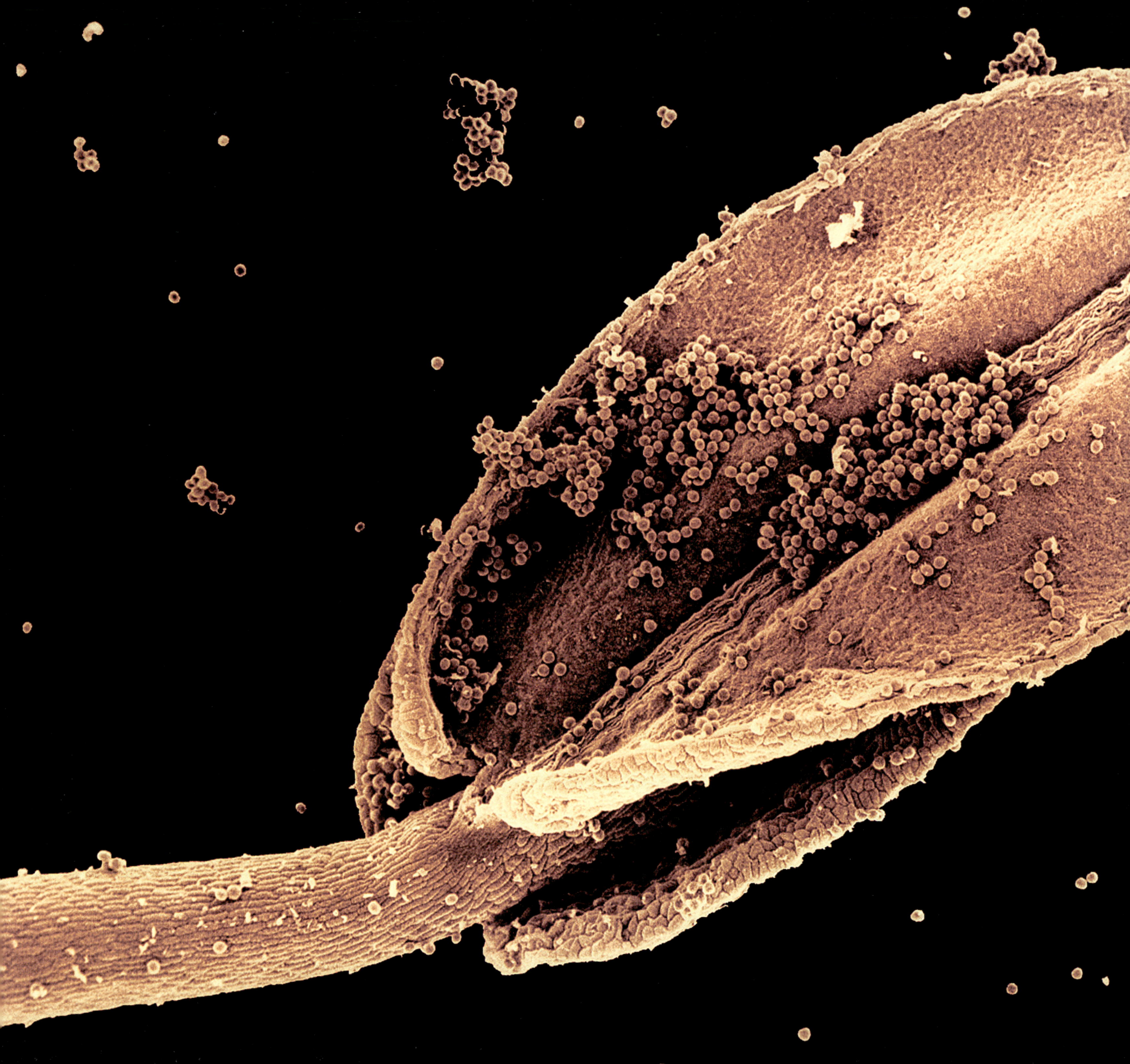

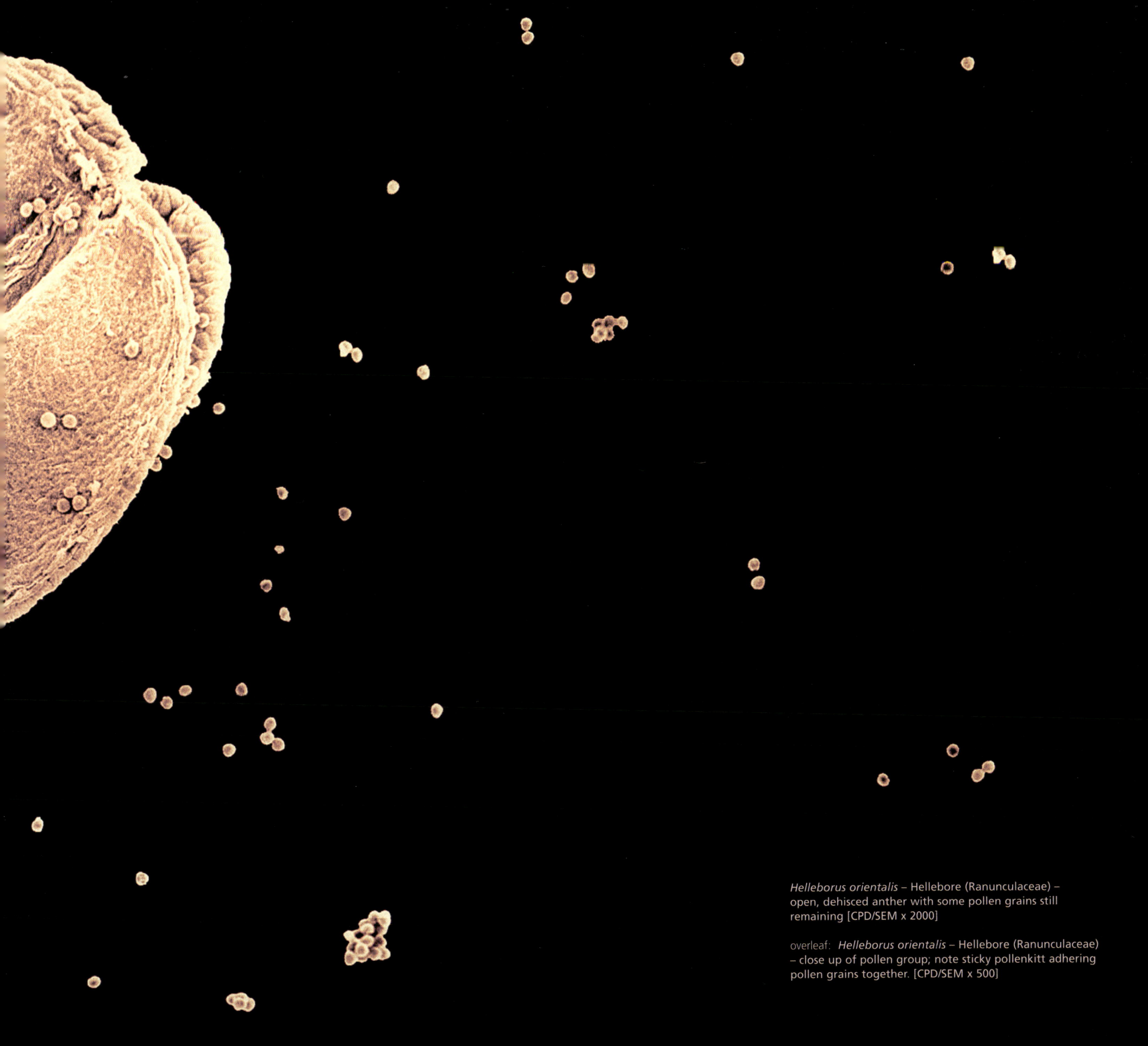

Helleborus orientalis – Hellebore (Ranunculaceae) – open, dehisced anther with some pollen grains still remaining [CPD/SEM x 2000]

overleaf: *Helleborus orientalis* – Hellebore (Ranunculaceae) – close up of pollen group; note sticky pollenkitt adhering pollen grains together. [CPD/SEM x 500]

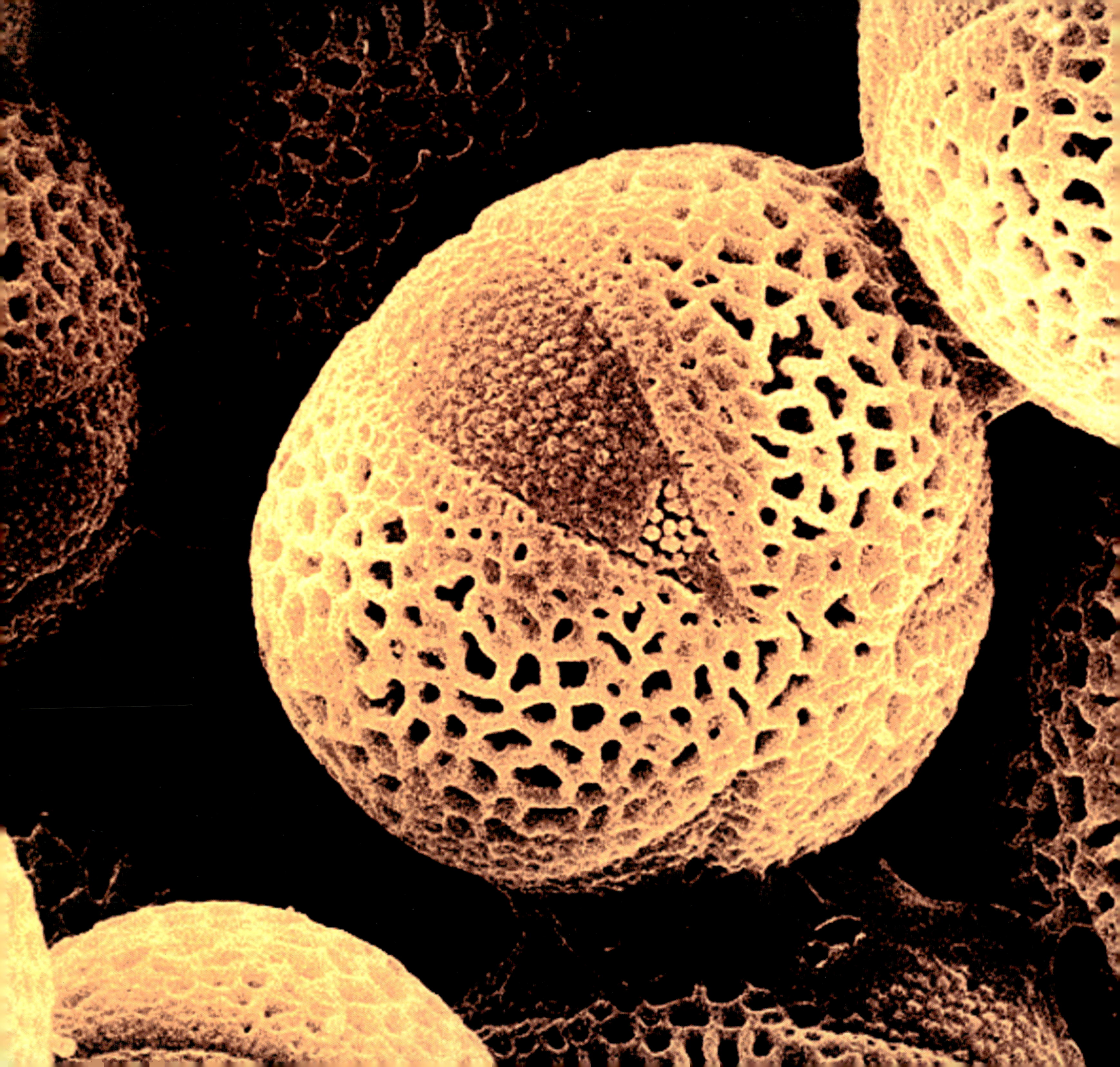

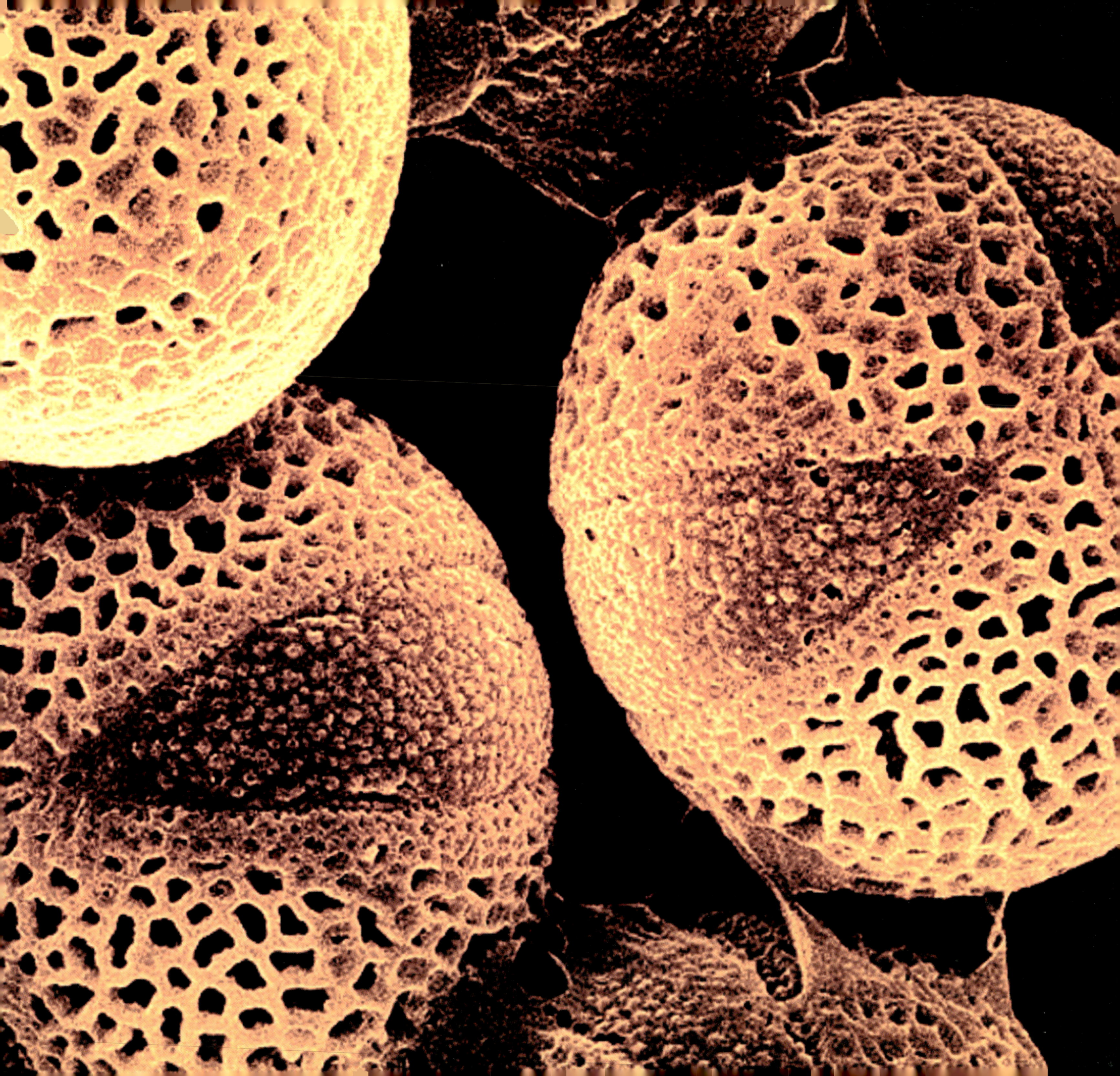

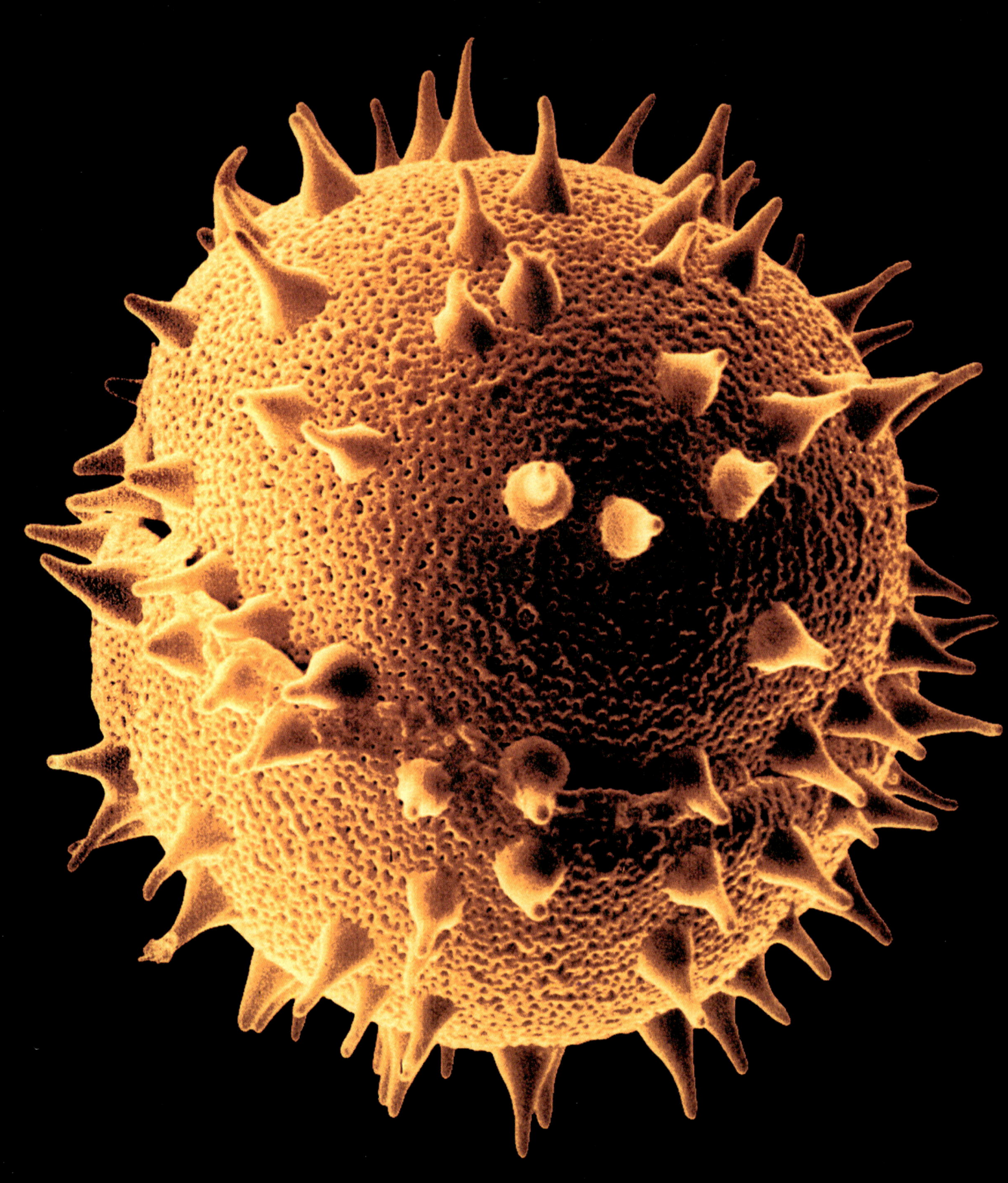

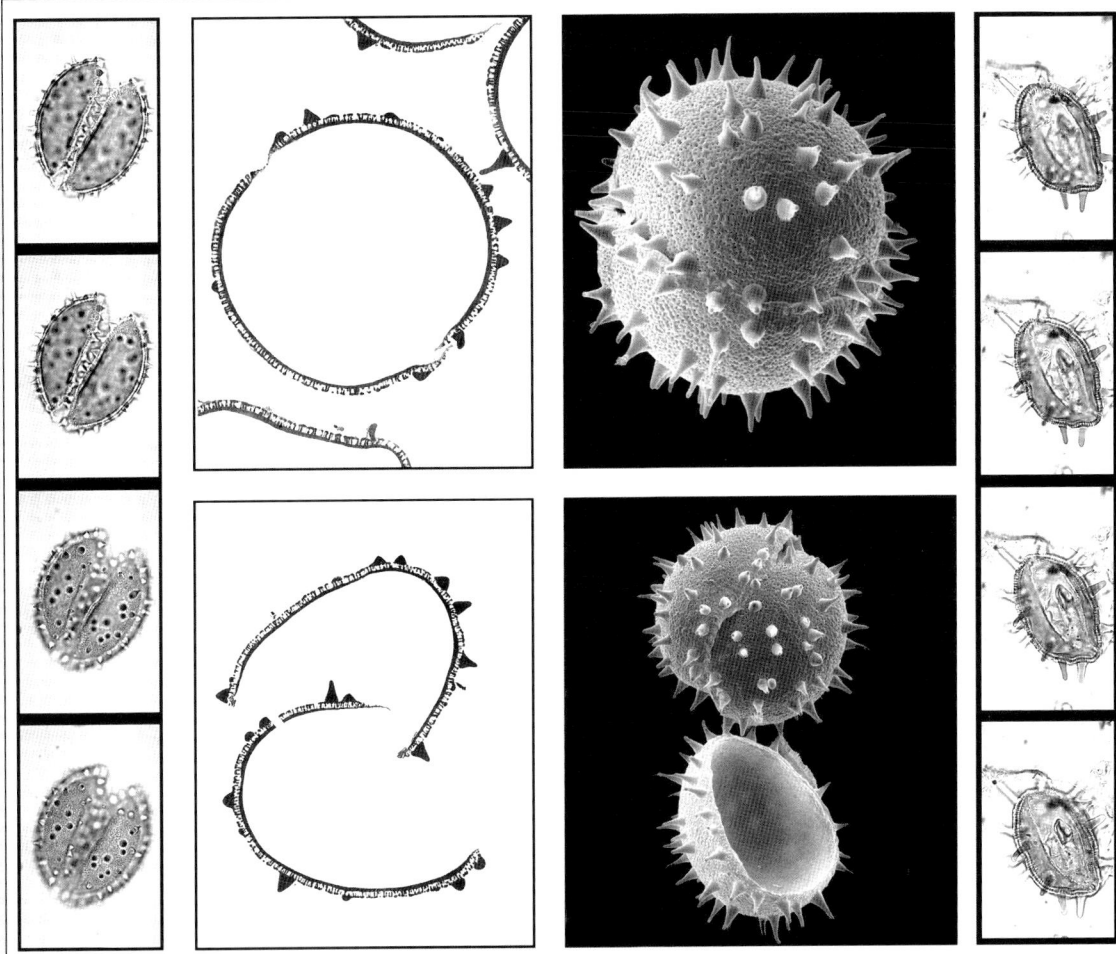

Nypa fruticans - Mangrove Palm (Arecaceae).
above: pollen grains in different views. When the pollen is acetolysed or fossilised the pollen usually separates into two halves, bottom centre – left and right: On the left a single pollen grain has been photographed from the light microscope at four different focal levels, note that each image highlights different details; on the right a single fossil pollen grain of a fossil Nypa palm has also been photographed at four differing focal levels in the light microscope. This is one of a number of fossil Nypa-like pollen grains recovered from the Isle of Wight London Clay deposits. It is about 55 million years old (Eocene). Its presence in the Isle of Wight deposits is indicative of a very much warmer tropical climate during the Eocene of NW Europe. Between the fossil specimens there is a range of different spine lengths, while the spines of living Nypa pollen are very uniform. The pollen spine data suggest that previously Nypa probably comprised more than the single species represented today. It was also very widespread throughout the palaeo-tropics, while today its natural distribution is confined to Malesia.
[LM x 700 – left and right; TEM x 1500 centre left; SEM x 1700 upper right centre, x 1000 lower right centre]
below: pollen details of supratectal spines and the exine wall [SEM x 10,000 – top left and bottom right; TEM x 10,000 – bottom left and top right – acetolysed]

opposite: *Nypa fruticans* – Mangrove Palm (Arecaceae) pollen grain, expanded condition – a ring-like aperture encircles the spiny spheroidal pollen.
[SEM x 1700 – acetolysed]

overleaf left: *Poa trivialis* Rough Meadow Grass (Poaceae) pollen grain, expanded condition [SEM x 2600]

overleaf right: *Phleum pratense* Timothy Grass (Poaceae) note exerted anthers to catch the wind

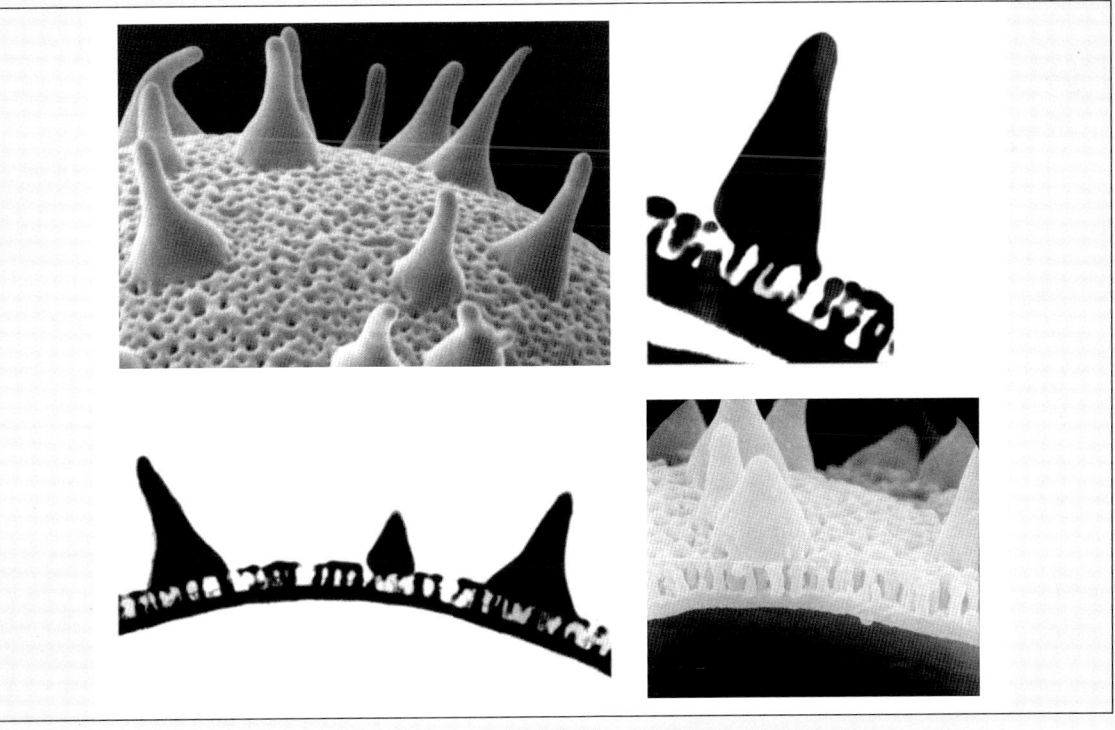

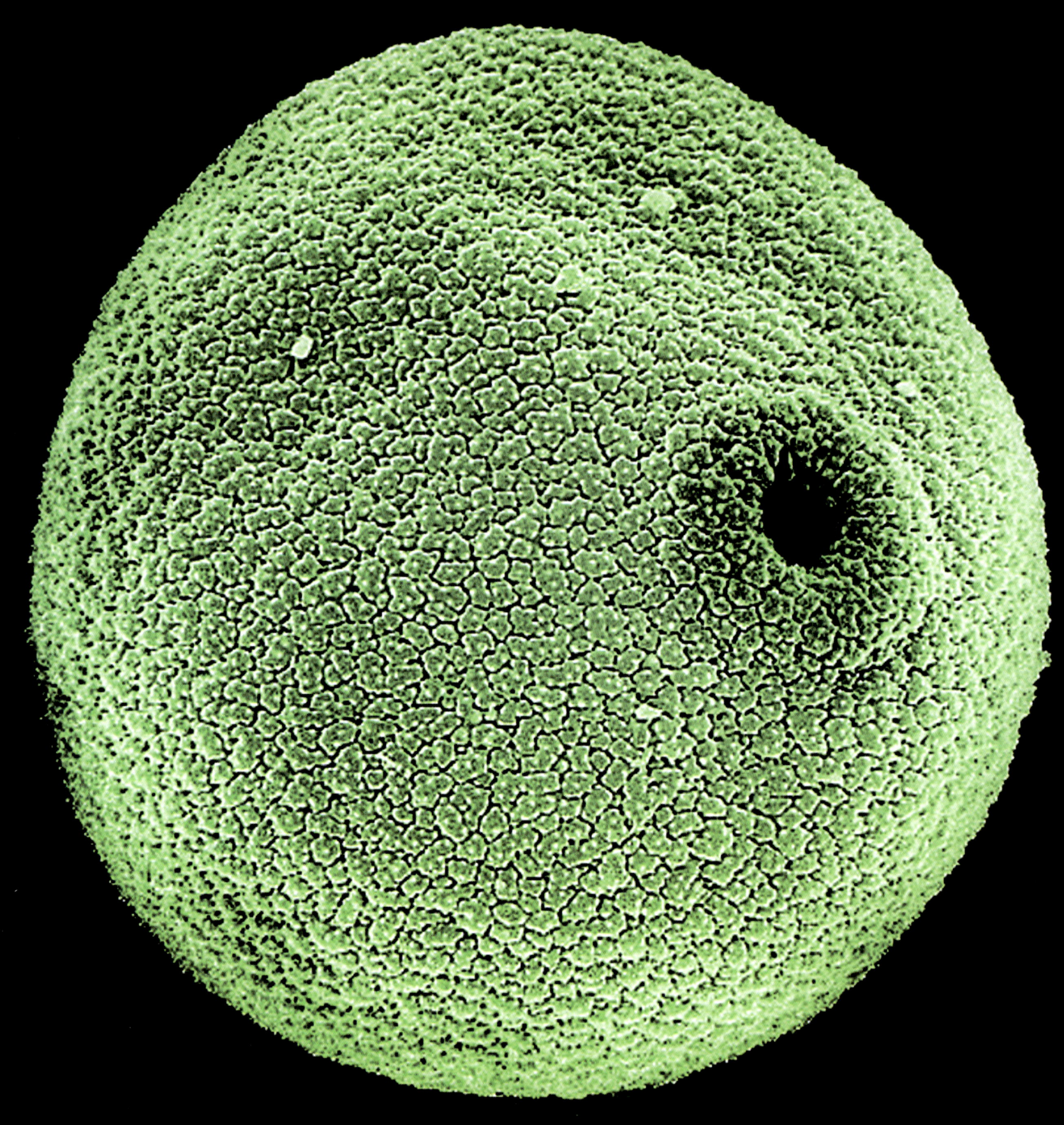

common metals in pollen, will also modify the absorption spectrum of flavonoids. To the human eye, pollen grains may contrast with or be similar in colour to the flower petals. Pale yellow and colourless pollen grains are common throughout the flowering plants. Pollen grains of other colours are usually associated with insect pollination, while wind or water-pollinated plants are usually pale or colourless.

Artificial pollen colour

If pollen grains are treated with a mixture of acetic anhydride and sulphuric acid ('acetolysed') – a standard laboratory preparation procedure for pollen, especially from dried herbarium specimens – or if they are preserved (fossilised) in rocks, pollen grains lose their external coating of lipids. They also lose the internal cytoplasm, including the organelles contained within it, and the non-sporopolleninous inner wall layer, the intine, and so are no longer viable. However, the sporopolleninous exine remains, and this is the layer that is used for pollen identification. Seen in light microscopy, fossilised or modern acetolysed pollen grains are light to rich golden brown in colour. In the laboratory pollen grains are acetolysed to reveal the often highly ornamented pollen wall because this is the part of the pollen grain which provides us with most of the characteristics we need for comparative pollen studies.

In the laboratory we can stain pollen grains variously to highlight different features. For example, we can contrast the sporopolleninous wall with the internal cellular material in order to check whether the pollen grains are likely to be viable. Biological studies of pollen development, mature pollen and/or pollen germination begin with fresh pollen, using fixation and staining procedures to arrest or highlight desired stages or attributes of the process(es) being studied.

Furthermore, the sophistication of modern computer technology allows us to colour pollen and other artefacts electronically, and there are many

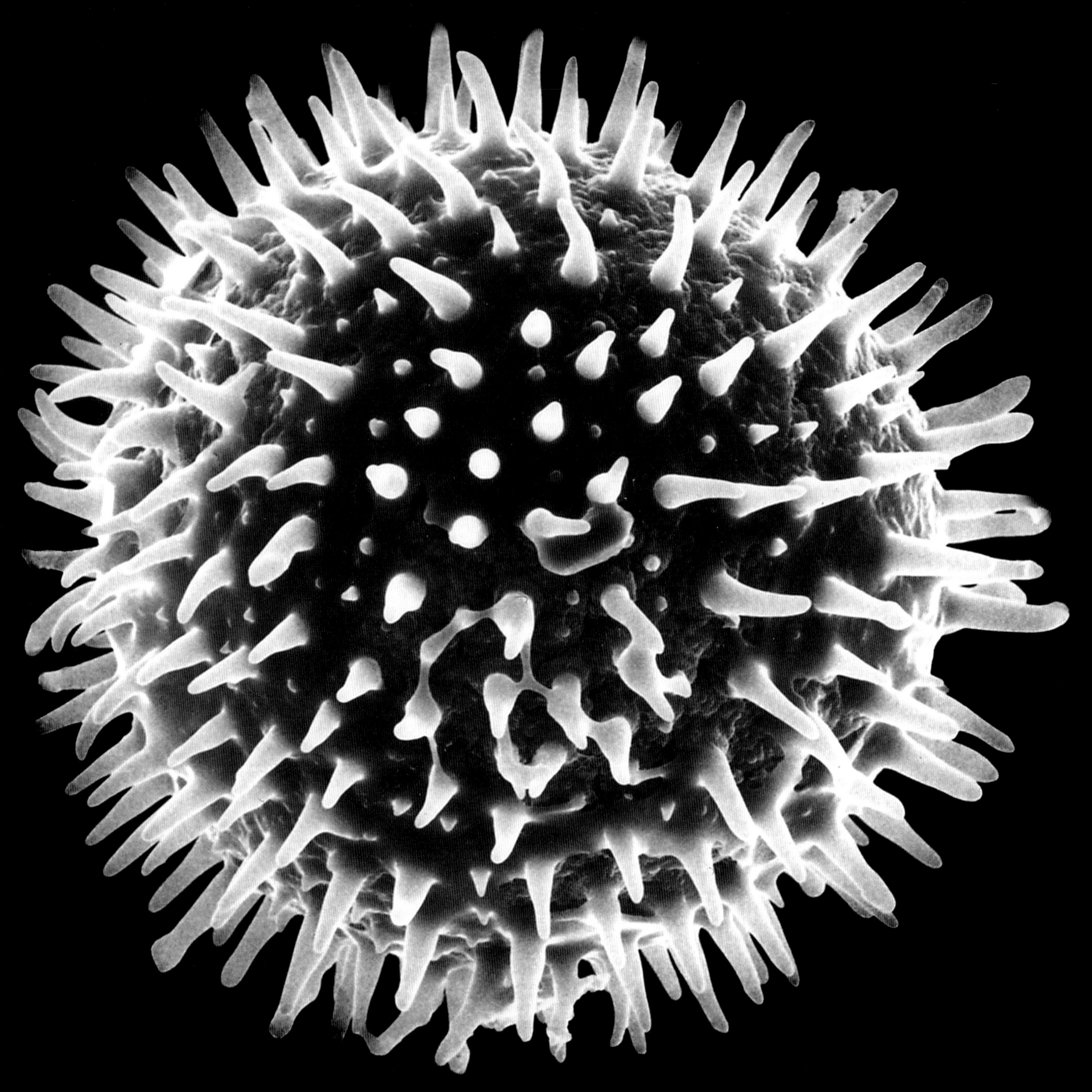

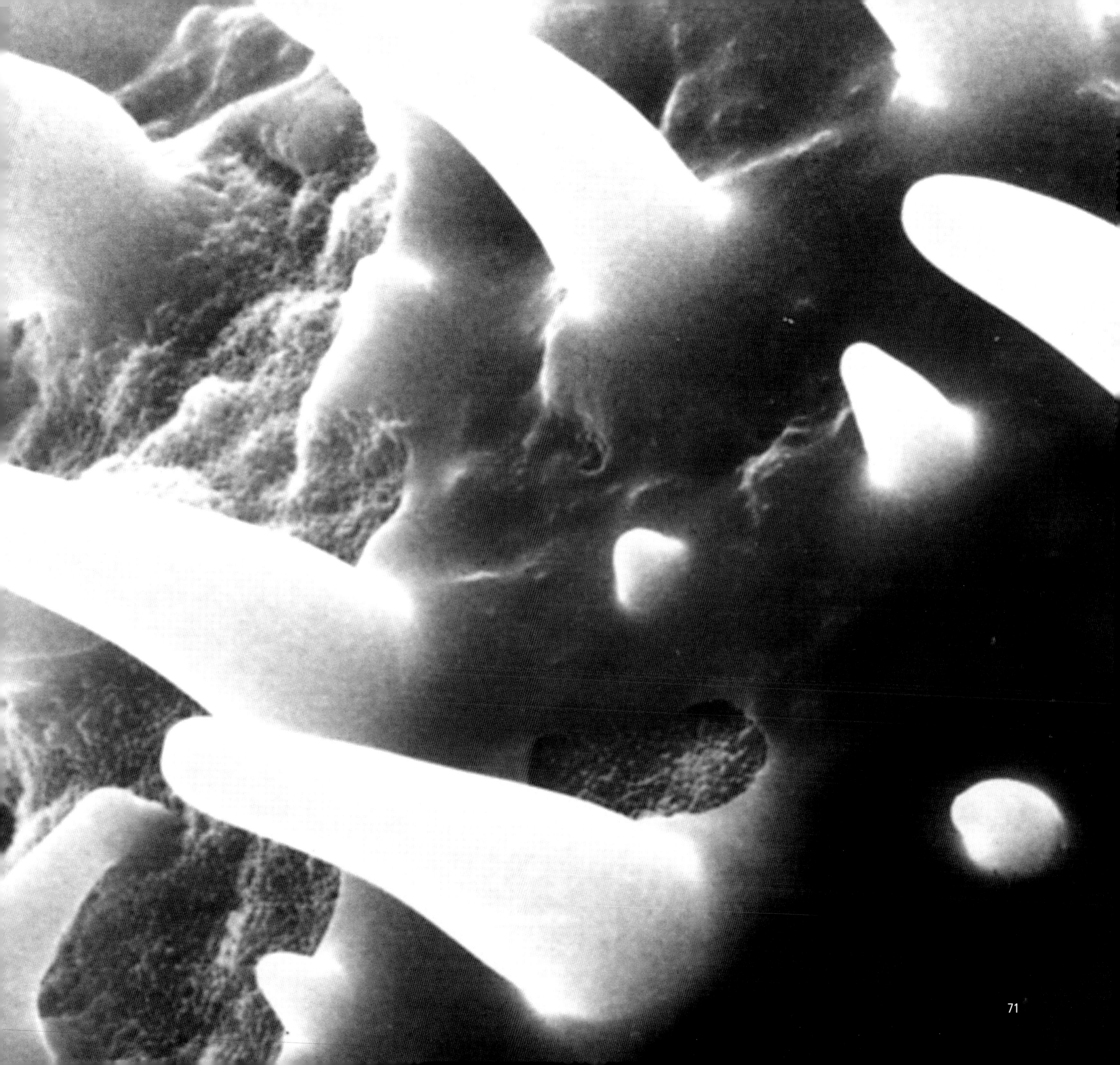

71

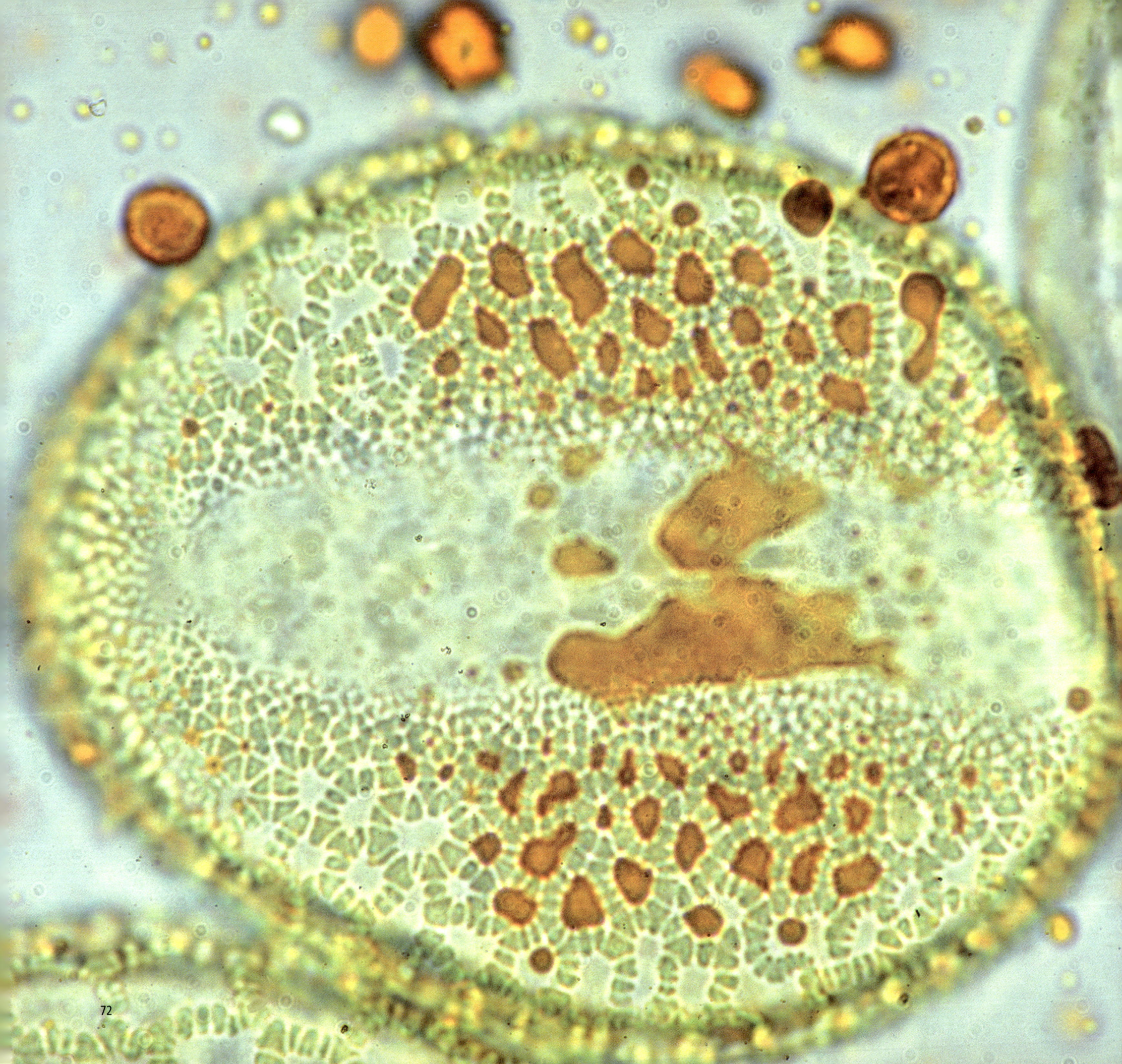

examples in this book. This can be very useful for highlighting features of pollen grain architecture, such as aperture areas, or special characteristics of the exine surface.

COMPARATIVE POLLEN STUDIES

Pollen has a number of qualities which, in combination, make it very useful in a range of comparative studies: it is very small, it has a highly distinctive exine with an enormous range of variations specific to particular groups and species of plants, and the exine is highly resistant to decay. Pollen grains, isolated from the plants that produced them, can frequently be identified to family level, often to generic level, and sometimes even to species level. Pollen and other airborne microscopic particles are collected on a daily basis in order to monitor levels of atmospheric pollution. The presence of pollen in soils, rocks, honey, clothes, faeces and so on, often helps provide answers in various types of investigation. In honey analysis the provenance of honey can be checked by its pollen content: Australian *Acacia* honey cannot masquerade as 'Scottish Heather' honey, without risk of detection, because the pollen types of *Acacia* and Heather are very different. In forensics, pollen sampling of soils, clothing or other articles related to the crime, can reveal time of year and place of a rape or murder, because the pollen present will reveal species of plants flowering at the scene of the crime. In cases of illegal crop spraying, pollen found on the corpses of bees will indicate the crops where the bees were foraging. Past vegetation and climate reconstructions rely very much on pollen data: for example we know that fifty-five million years ago mangrove palms flourished on a tropical Isle of Wight. We also learn a great deal about the evolution of flowering and non-flowering plants from studying the fossil pollen record combined with records of larger fossilised plant organs such as leaves and fruits; for example, the earliest occurrences and distribution of tree ferns, conifers, and early flowering plants such as the *Magnolias*, in the archaic world of

previous page left: Abutilon pictum (Malvaceae) – a pollen grain, expanded condition. [CPD/SEM x 600]

previous page right: Abutilon pictum (Malvaceae) – close up of pollen exine showing oily lipid (pollenkitt) between spines. [CPD/SEM x 2000]

below: Lilium tigrinum – Tiger Lily (Liliaceae) – pollen grain exuding copious lipid (pollenkitt) from the net-like exine. (Francis Bauer, between 1790 and 1840)

opposite: Hemerocallis cv. – Day Lily (Hemerocallidaceae) – pollen grain showing distribution of oily lipid in the exine network (*Lilium* and *Hemerocallis* have very similar pollen). [LM x 40]

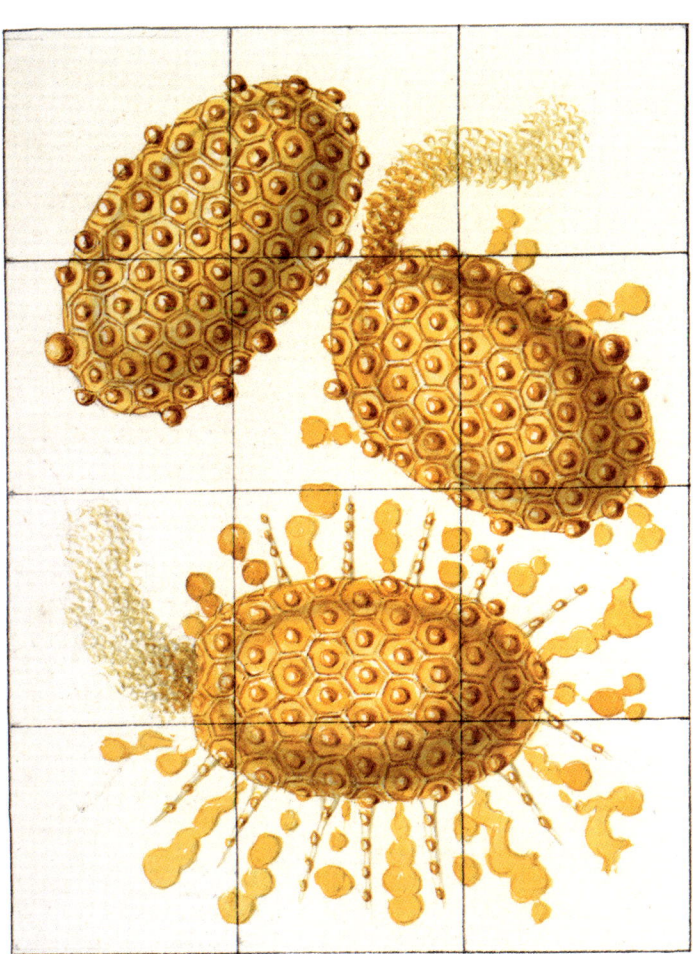

Laurasia and Gondwana. Many of us are now familiar, from radio, television and films, with the contribution of pollen studies to archaeological reconstructions of diet and agriculture through the study of pollen in pots, middens, faeces, and soils around settlements.

POLLEN MORPHOLOGY IN PLANT SYSTEMATICS AND EVOLUTION

In most of the examples of comparative pollen studies mentioned above the pollen, whether it is fossilised, or from a murder enquiry, honey sample, archaeological site, air monitoring sample or bee poisoning, involves a straightforward comparison between the sample and a reference collection of microscope slide preparations. A reference collection may be a vast collection of microscope slides covering pollen from thousands of species of plants, or highly specific, depending on the nature of the investigation. However, one study area, pollen 'morphology' in relation to plant systematics and evolution, is more complex in terms of the comparisons it makes. It seeks not only to describe the variation in pollen from the species of plants being studied, but also to use the resulting data to contribute to a better understanding of the relationships that exist between the plants studied and other plant groups. Comparisons are also made with the pollen fossil record, to find earliest occurrences and palaeogeographic distribution of similar pollen types, and to try and understand how the plants that produced them evolved and diversified over time. The reason that fossil pollen grains are so useful in reconstructions of plant evolution is that they are rather conservative with respect to changing their appearance over many millions of years. Unlike the whole plant, they are not exposed to the elements during their life and unlike the whole plant they do not have to adapt to local conditions, or to changes in climate and altitude, in order to survive; they are snugly protected inside the plant until released into the atmosphere at maturity.

opposite: *Pavonia urens* (Malvaceae) – a pollen grain [LM x 400 – natural colour]

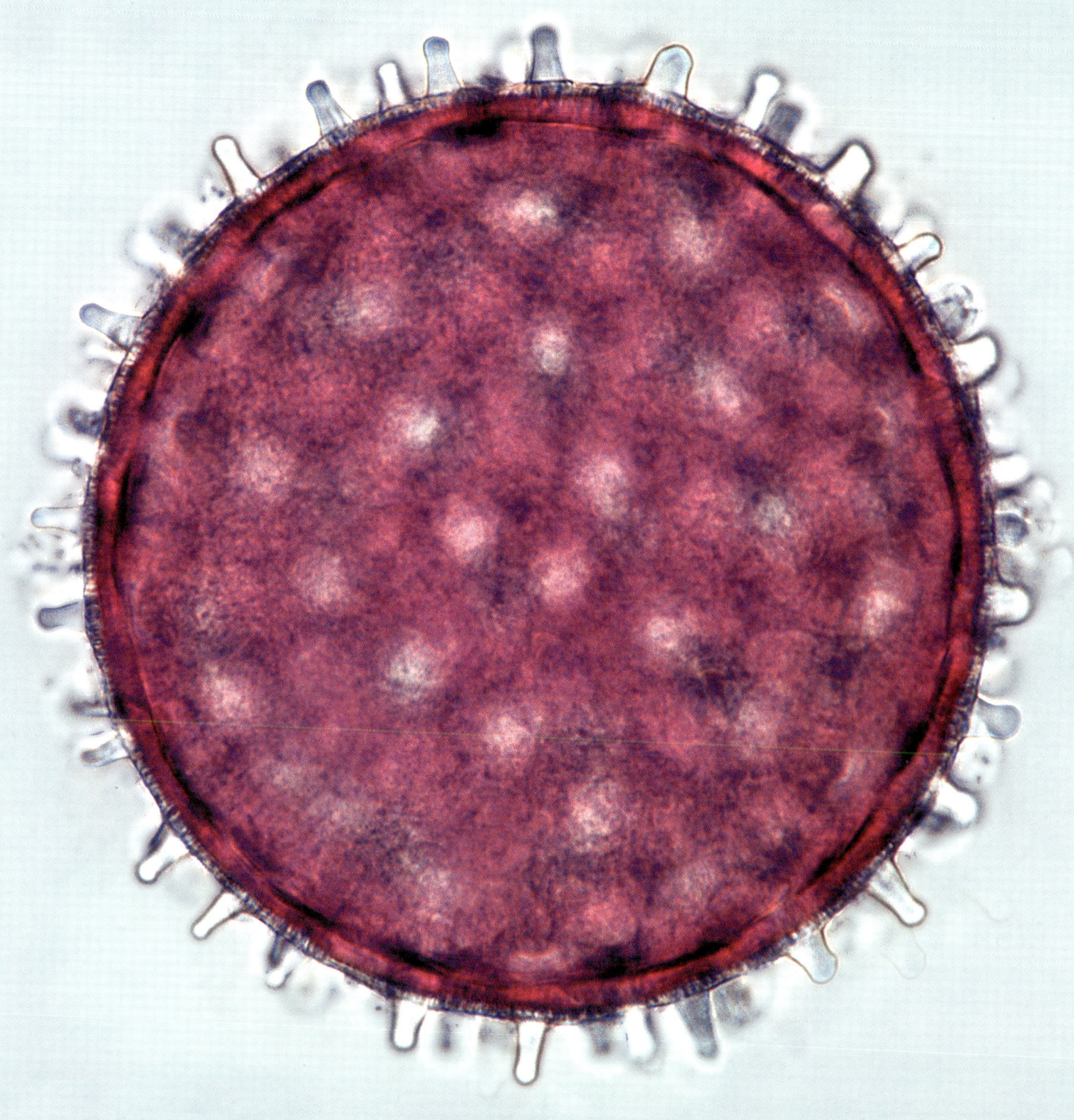

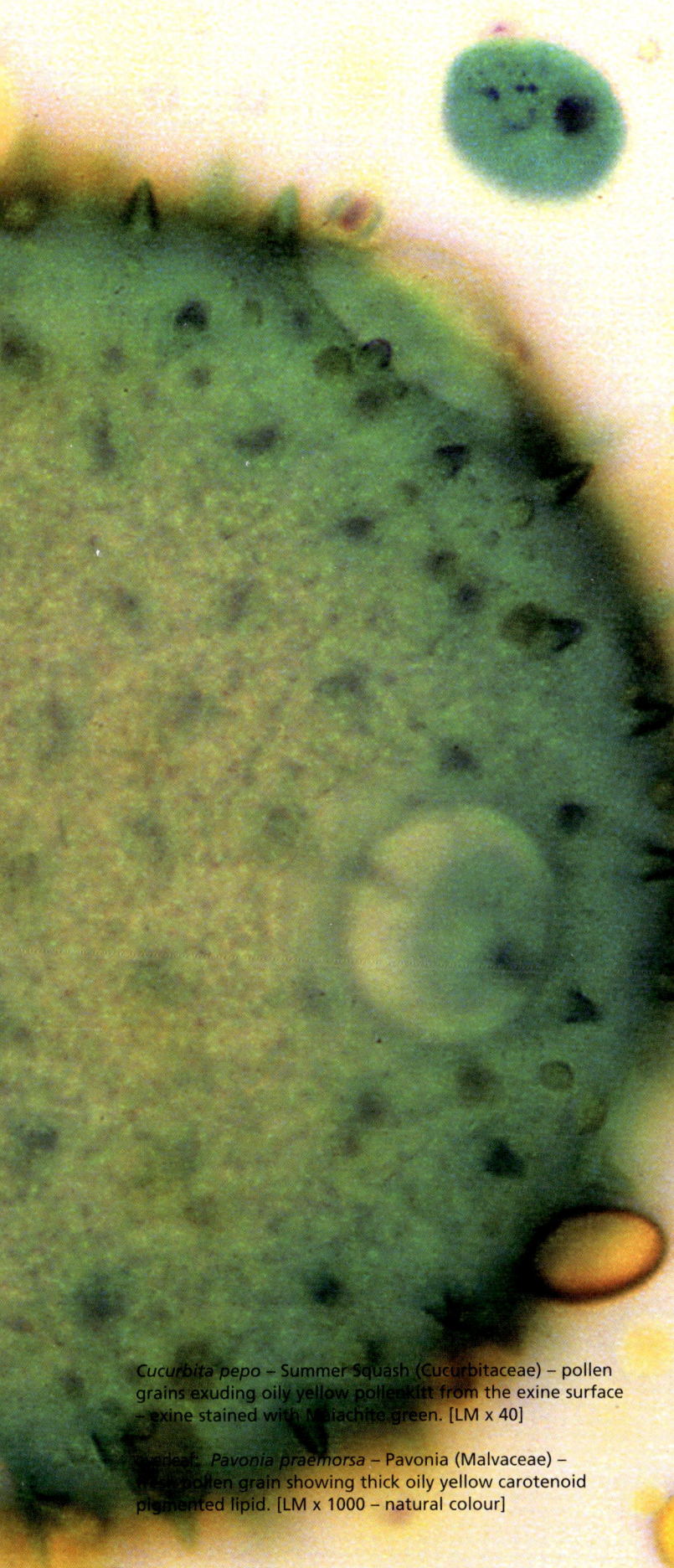

Cucurbita pepo – Summer Squash (Cucurbitaceae) – pollen grains exuding oily yellow pollenkitt from the exine surface – exine stained with Malachite green. [LM x 40]

Opposite: Pavonia praemorsa – Pavonia (Malvaceae) – fresh pollen grain showing thick oily yellow carotenoid pigmented lipid. [LM x 1000 – natural colour]

Pollen studies in plant systematics and evolution are not undertaken in isolation but in collaboration with other research botanists, working on the same group(s) of plants. They will be studying other regions of the plant including anatomy, embryology, floral development, cytogenetics, DNA, bio-chemistry, physiology, biogeography, palaeobotany and so on. Pooling the results allows us to develop increasingly refined ideas/hypotheses about how and when different groups of flowering plants evolved, and their relationships to each other ('phylogenetics').

A simple analogy might be trying to unravel a very complex family tree which has a large amount of information about relationships already recorded by previous members of the family, some right, some wrong, some pure tittle tattle, and much missing (including a few illegitimate children that the ancestors preferred to overlook). The keen historian will also develop an informative database from a variety of sources including local archives, church records, books, photographs, letters, postcards, diaries, as well as conversations with elderly relatives and other interested parties. The main differences between the family tree, and a phylogenetic tree of, for example, flowering plants is neither the research approach, nor the fact that human beings are phenomenally interested in themselves and love constructing family trees. It is the time scale and the number of species involved. Human beings are just one species, *Homo sapiens* L. which, over three or four million years, has evolved from ancestors with more than a passing resemblance to chimpanzees; while flowering plants, with more than 250,000 species, have evolved over at least 120 million years. So botanists are working, not only at a very different level in terms of species numbers, but also of time scale.

Today DNA has a central role in phylogenetic studies of all living organisms – DNA codings remain true within an organism, its 'blueprints'; they are non-adaptive and uninfluenced by external pressures such as environment or climate. DNA is already being used to trace the evolution of

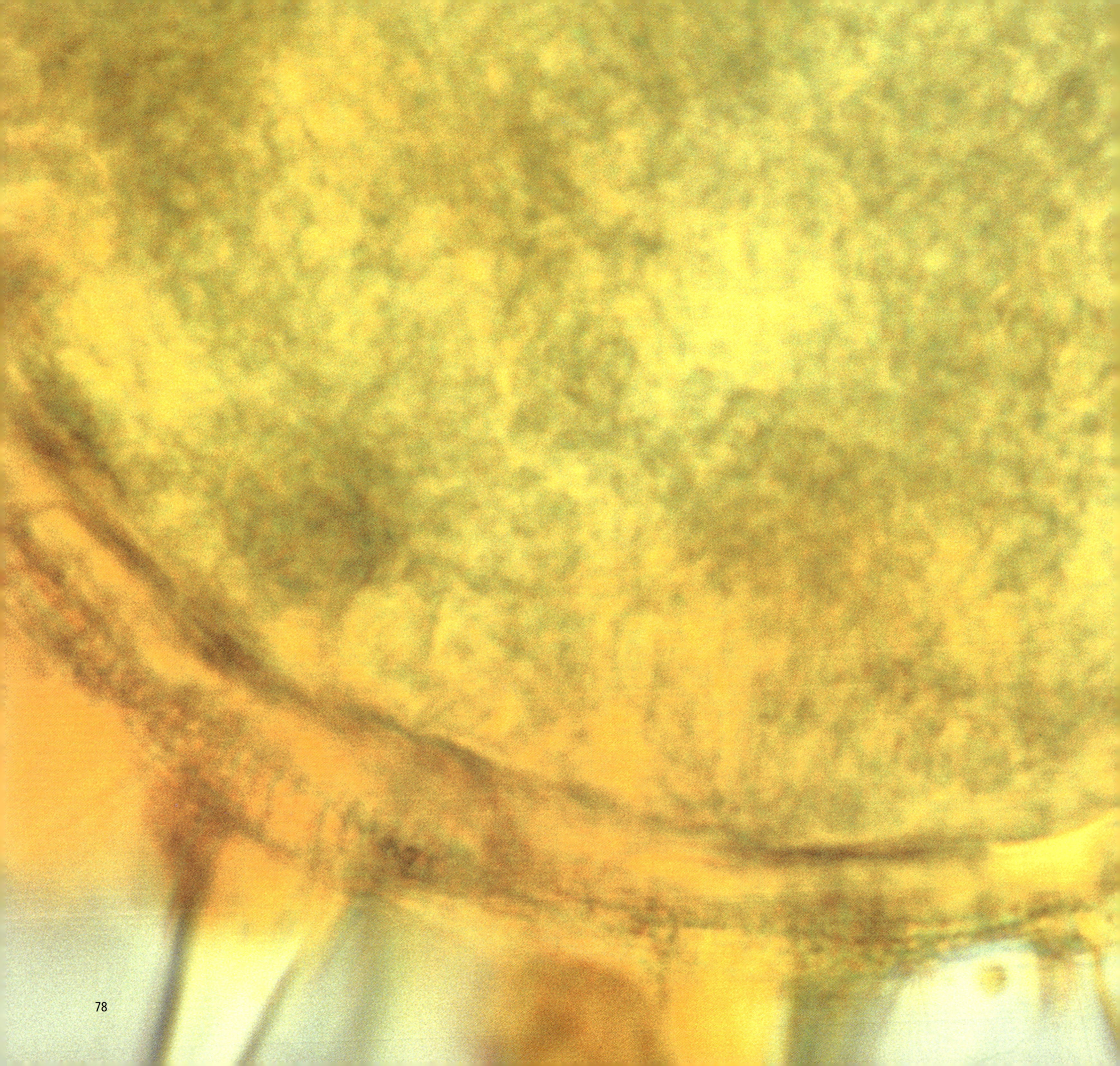

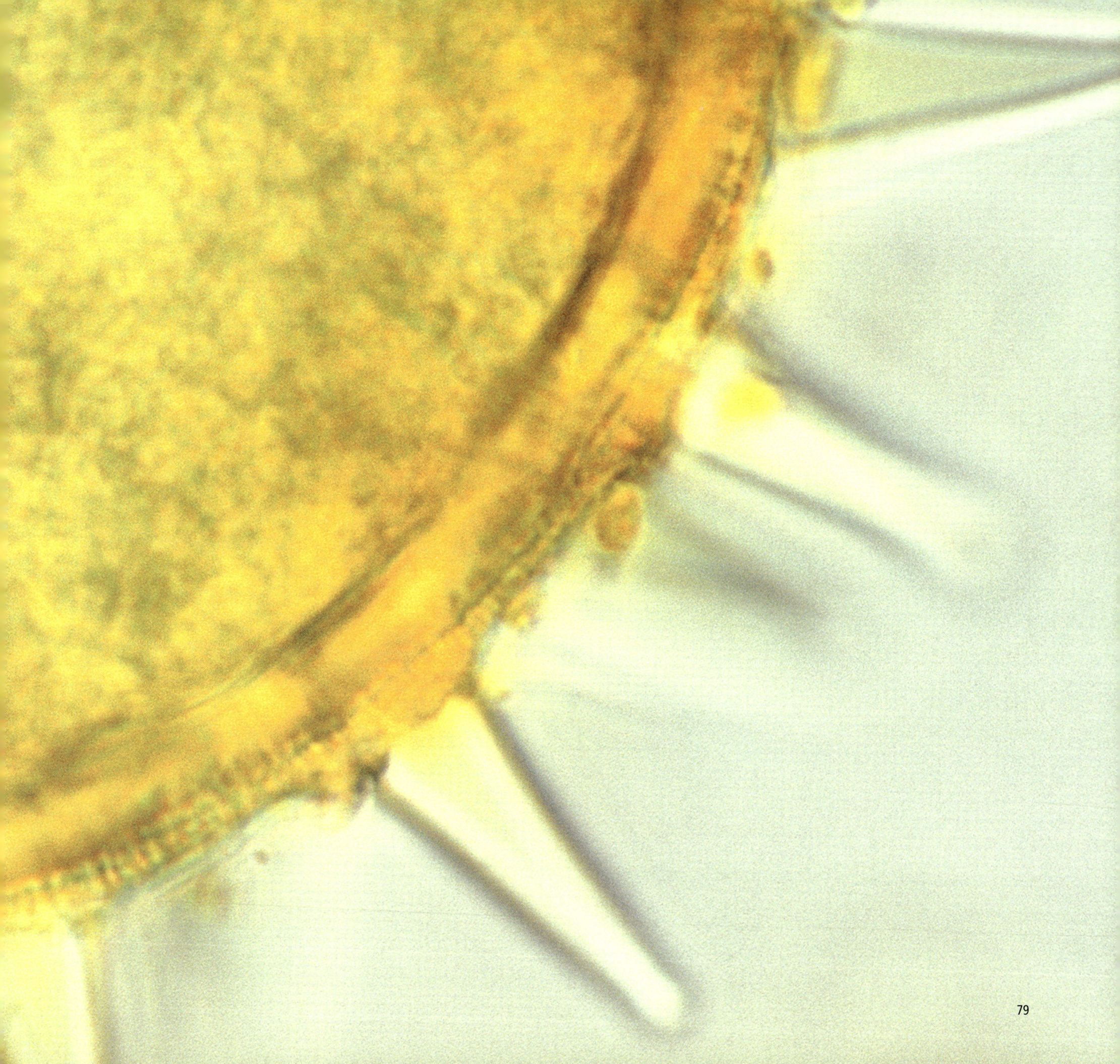

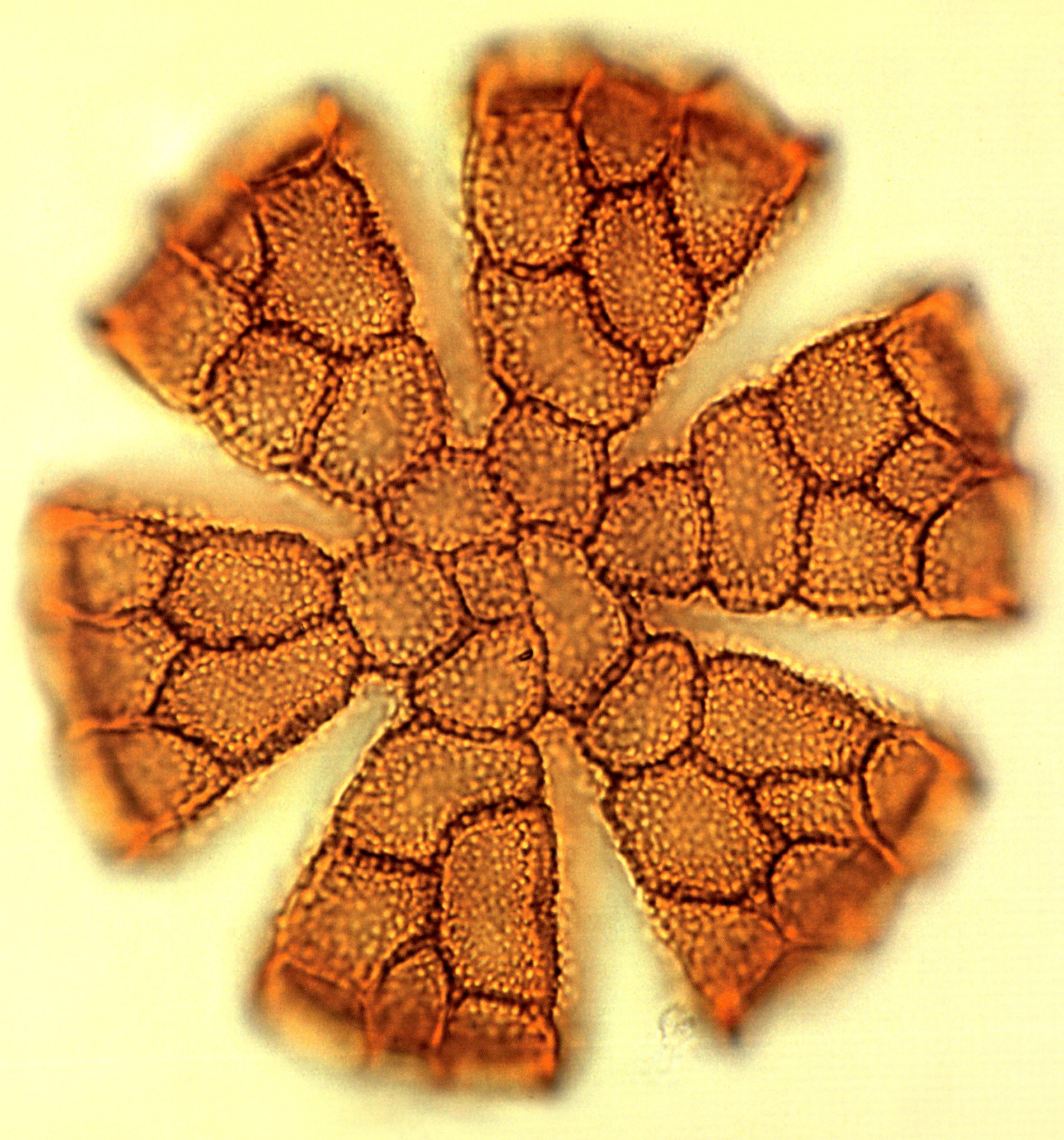

80

different races of *Homo sapiens* and, over time, will no doubt figure increasingly, in research into family trees. Genetic codings of DNA for different species have also taken a central position in studies of plant relationships, usually combined with morphological and other data, in order to carry out more rigorous phylogenetic analyses. It is interesting to note that, although pollen grains are huge compared with molecules, pollen morphological data often reflects results from molecular data. This is probably largely due to the morphological conservatism of pollen over excessively long periods of time.

POLLEN PREPARATION FOR COMPARATIVE POLLEN MORPHOLOGY

Detailed data about pollen characteristics are needed for comparative studies of pollen morphology. The raw pollen material may be fresh, in which case it is often prepared by critical point drying (CPD), a procedure that takes the pollen through liquid CO_2 to a dry gas phase, to prevent the pollen grain from collapse and to preserve other features in their natural state, for example pollenkitt. Pollen, because of the richness of the collections necessary in a systematic study, is more frequently taken from an herbarium – a collection of pressed and dried plants for scientific study. There is a range of drying, and fixation techniques used to prepare fresh pollen for microscopy, while pollen from herbarium specimens is usually acetolysed. Subsequent examination of the prepared pollen material is carried out using a combination of light microscopy, scanning electron microscopy and transmission electron microscopy. Each method allows certain features of the pollen grain to be seen more clearly than either of the other two methods allows. Although the highest resolving power (c. 1000-1500x) of a light microscope is much, much lower than that of either a scanning electron microscope or a transmission electron microscope – both of which have a resolving power for biological material of about 50,000-60,000x – light microscopy has particular advantages

opposite: *Ocimum kilimandscharicum* – a Basil relative (Lamiaceae) – pollen grain focussed to show surface pattern of exine. [LM x 400 – acetolysed]

overleaf: *Trachycarpus fortunei* – Chusan Palm (Arecaceae). left: Palm 'tree' in flower – note huge inflorescences; centre: close up of the inflorescence – note anthers emerging from masses of tiny flowers; right: *Chamadorea elegans* – Parlour Palm (Arecaceae) pollen grain – with a single aperture and a network-like (reticulate) tectum.
[SEM x 4000 – acetolysed]

page 84: *Ilex aquifolium* – Holly (Aquifoliaceae).
above: a fractured pollen grain to show wall structure;
below: a whole pollen grain in polar view [SEM x 3000 – acetolysed; bottom: whole pollen grain polar view, to show ectexine structures. [SEM x 2000 – acetolysed]

page 85: *Ilex aquifolium* – Holly (Aquifoliaceae) – whole pollen grain equatorial view, to show ectexine structures. [SEM x 2000 – acetolysed]

82

No Flowers – No Pollen; No Pollen – No Flowers

over both scanning and transmission electron microscopy. It is by far the most accurate method for measuring the various dimensions of pollen grains, such as overall size and shape, aperture size and shape, as well as wall thickness. It is also the essential tool for routine searching and examination of dispersed pollen grains, such as those found in rock, soil or honey samples. In spite of its comparatively low resolution the light microscope, used alone, provides a body of information than neither the scanning nor transmission electron microscope, in spite of their high resolution, can match single-handedly. This is because in light microscopy, light rays are transmitted through the subject and allow it to be viewed at different focal levels, from the upper surface to the lower surface. As we focus through the object, the patterns which appear light on the upper surface become dark on the lower surface. This type of analytical microscopy is termed, 'LO analysis' from the Latin *lux-obscuritas* (light-darkness). The beam of electrons utilised in scanning and transmission electron microscopy effectively focuses only on the surface of the object being viewed. A scanning electron microscope produces visually appealing images which capture the three-dimensional reality of the subject being examined. Furthermore, it is invaluable for resolving very fine detail, such as exine surface, apertures, and supratectal structures, which cannot be satisfactorily resolved with light microscopy. Pollen grains can also be fractured, so that the details of the internal structure of the walls and apertures can be observed. However, it is only by putting the pollen grains into epoxy resin, hardening the resin and cutting ultrathin sections through the block of pollen-dense resin, that the fine structure ('ultrastructure') of the pollen walls and apertures can be observed in detail. There are so many variants on basic pollen wall ultrastructure that this information can be of immense value in all branches of comparative pollen morphology, but especially in plant systematics and fossil pollen studies. To cut slices through the resin a very small, precision-made glass or industrial diamond knife, is used. The knife is fitted into a highly specialised

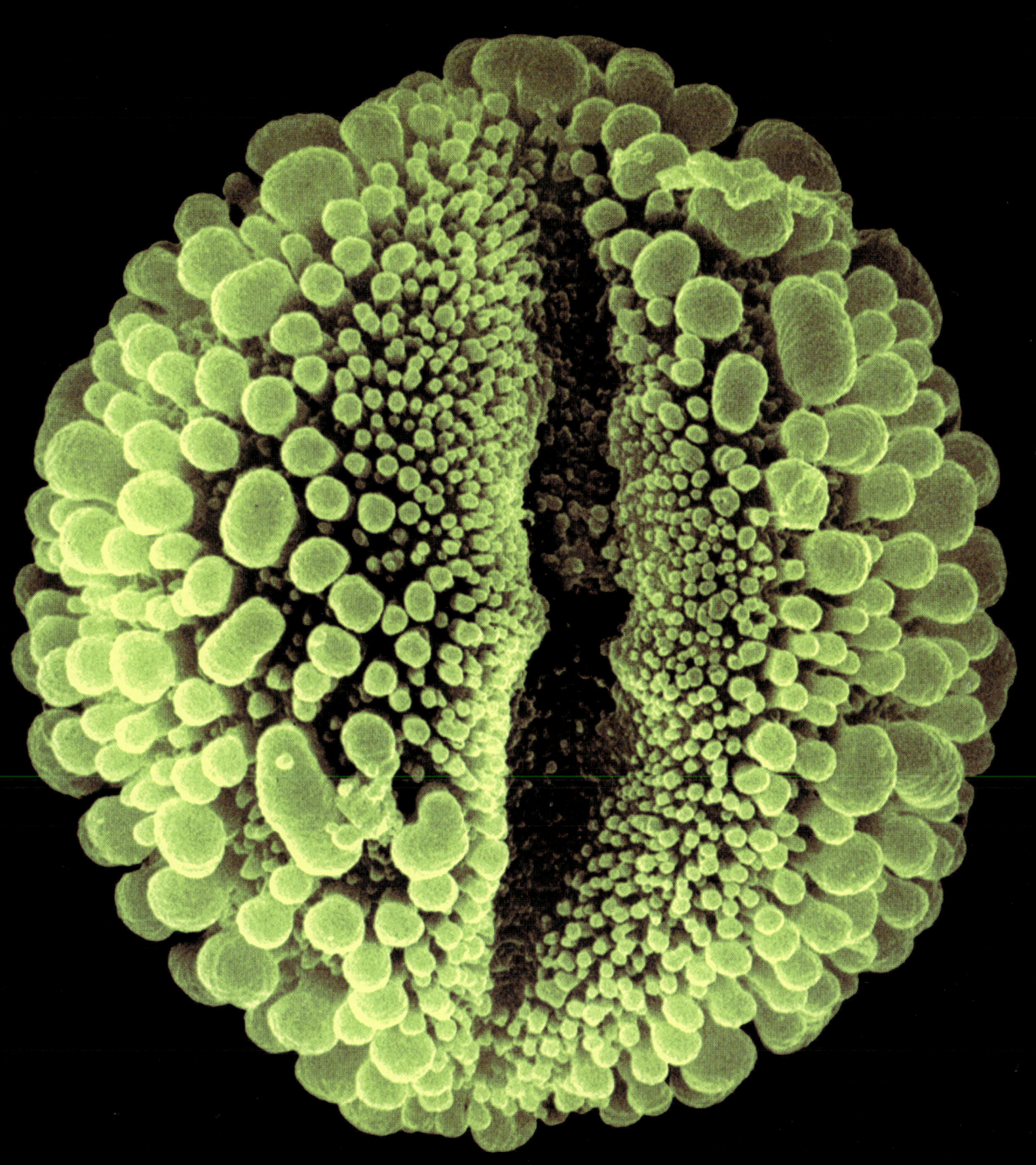

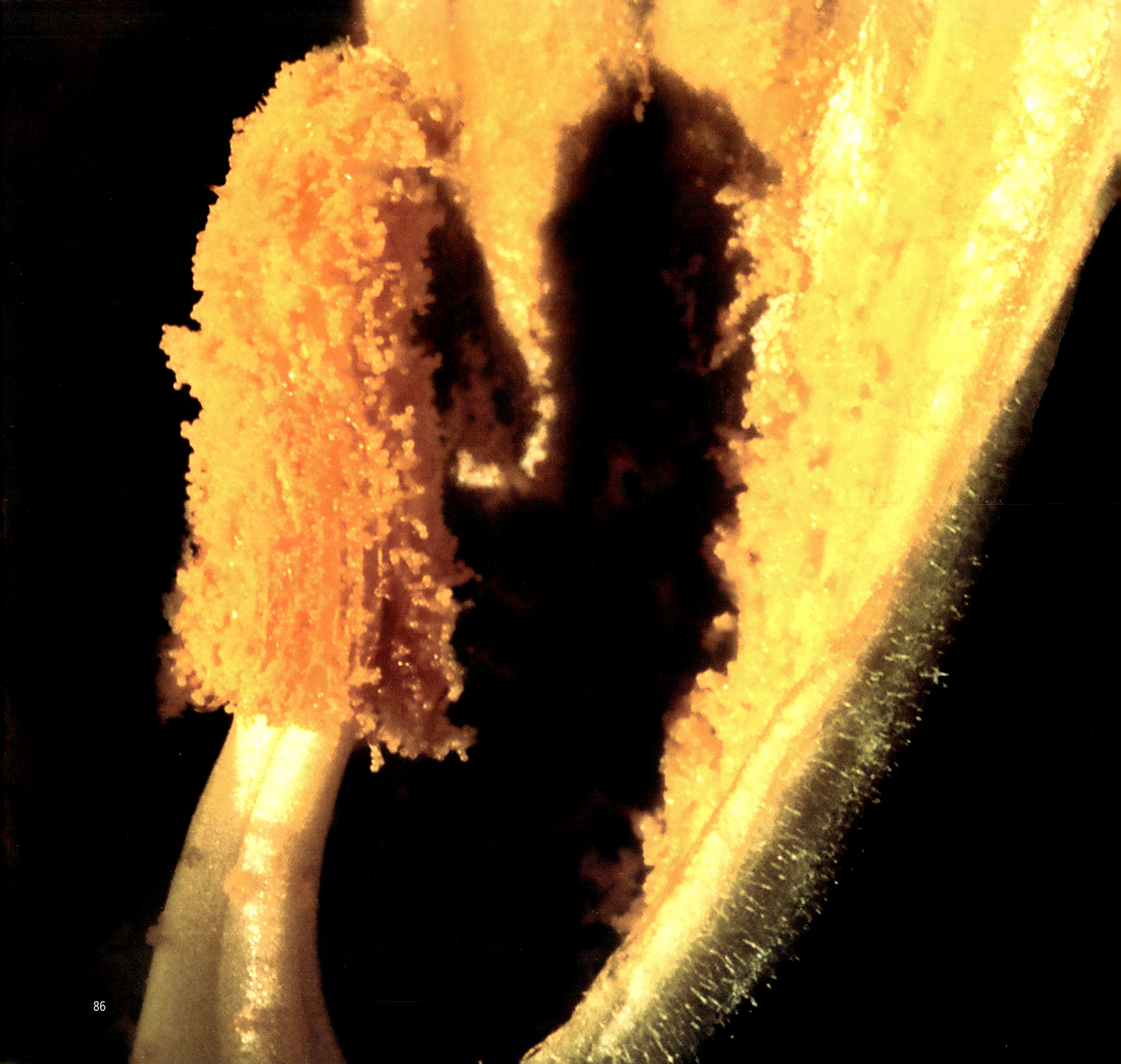

slicing machine, an 'ultramicrotome'. The resultant slices of pollen-filled resin, which are c. 60-100 'nanometres' (one thousandth of a micron) thick, are stained with special stains developed for resin-embedded biological material. Subsequently they are examined with a transmission electron microscope.

SEXUAL PLANT REPRODUCTION

Why is it necessary for plants to have such an elaborate structure as the pollen wall to contain their sperm cells when the sperm cells of animals are naked and they procreate successfully? The reason is simple: animal sperm cells pass from male to female in a moist environment. In their brief journey they are never exposed to the vicissitudes of the outside world but are still able to achieve cross-fertilisation for the health and vigour of the species. Plants are unable to do this because they cannot move about; they do not have legs or wings; they are rooted, quite literally, to the spot. This introduces a technical problem: how to achieve cross-fertilisation without having the sperm dry up *en route* to the waiting female. Clearly the answer is to put it in an airtight container. However, when it reaches the female of the species, the sperm has to get out of the container, and achieve its goal of fertilisation quickly, while it is viable. Not only must the container be airtight, to protect the sperm cells from dessication during transit, it must also have at least one opening to allow exit of the pollen tube carrying the sperm. For successful pollen germination the container must be capable of absorbing moisture, allowing expansion of the intine and bursting. The pollen exine is remarkably functional: the walls and, in many instances, the pollenkitt, provide protection from dessication and even solar radiation during transit; the aperture(s) allow for intine expansion and the aperture membranes, which are much thinner than the pollen walls, are designed to burst as the intine expands, so that the pollen tube can germinate.

Encasing the sperm cells in a tough sporopolleninous wall for protection is one part of the successful sexual strategy of plants. The other requirement, to

previous page left: *Cucurbita pepo* – 'Patty Pan' squash (Cucurbitaceae) – male flower, dehiscing anther.

previous page right: *Cucurbita pepo* – 'Patty Pan' squash (Cucurbitaceae) – female flower; note insignificant, undeveloped, staminal ring surrounding base of style.

opposite: *Allium fistulosum* – a species of Onion (Alliaceae) – the inflorescence opening from its thin papery spathe/bract.

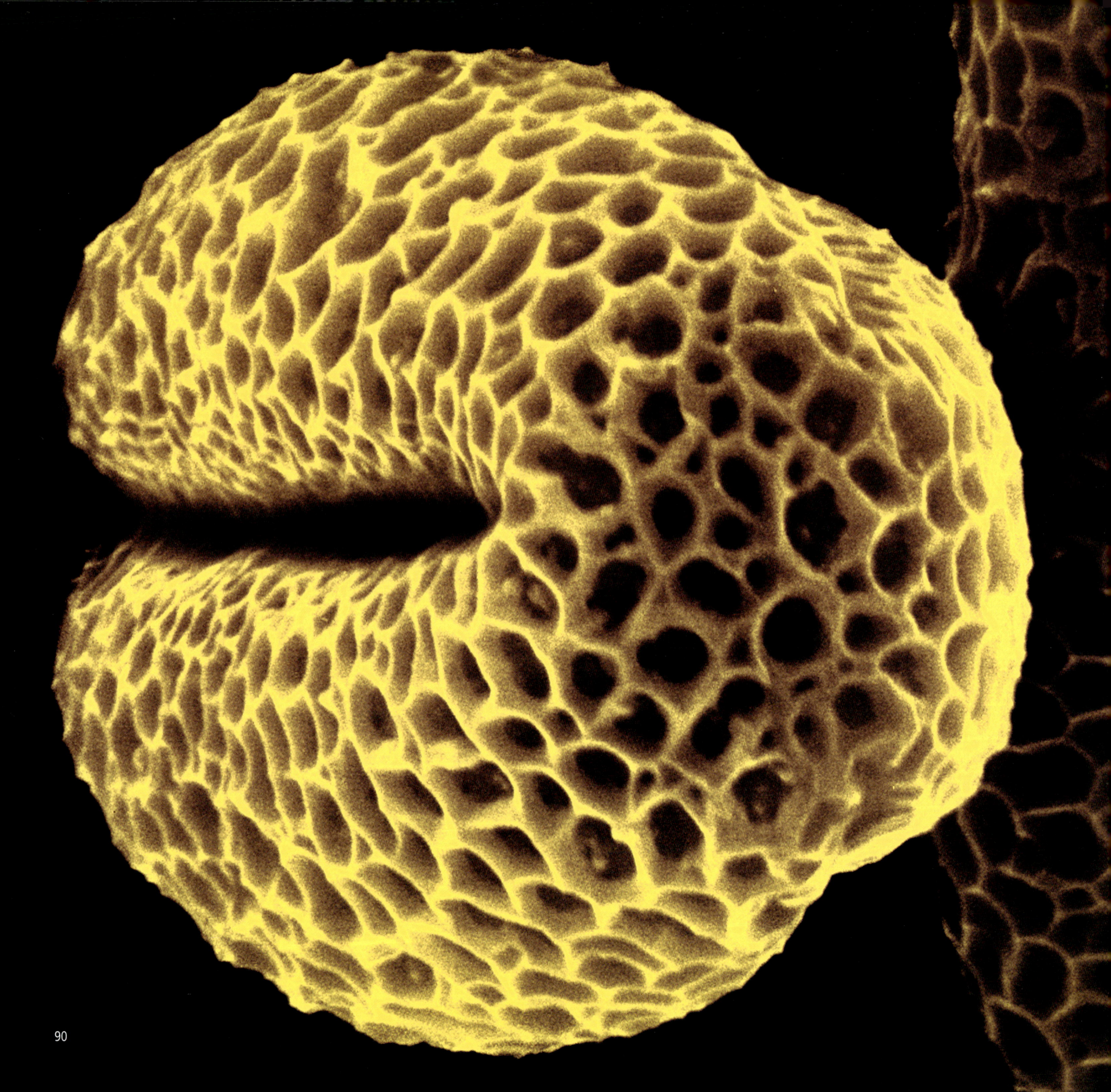

90

avoid inbreeding, is to transport the pollen grains carrying the sperm to the receptive (female) stigmatic surface of another plant of the same species. The main carriers ('pollinators') include wind, water, insects, birds, and small mammals (including bats).

The knowledge that plants are able to reproduce sexually goes back a very long way. Certainly the Ancient Assyrians (800BC) understood that flowers needed pollen to be passed from the male to the female organ to be pollinated, so that fruit would develop. There are a number of carved stone reliefs, some now in museums (including the British Museum, London) depicting pollen being transferred from male to the female inflorescences, apparently of date palms. However, sexuality in plants does occasionally still come as a surprise to some people; although it is doubtful, in modern society, if there are many people who would find it shocking. However, when the poet and botanist, Johann Goethe was writing in the late eighteenth century things were a little different:

"For the instruction of young persons and ladies this new pollination theory will be extremely welcome and suitable. In the past the teacher of botany has been placed in a most embarrassing position, and when innocent young souls took text book in hand to advance their studies in private, they were unable to conceal their outraged moral feelings. Eternal nuptials going on and on, with the monogamy basic to our morals, laws, and religion disintegrating into loose concupiscence – these must remain forever intolerable to the pure-minded."

What are flowering plants?

Flowering plants are everywhere in our lives, serving our everyday basic needs, glorifying the parks and gardens we walk in. We use them symbolically in bridal bouquets for joy, and in funeral wreaths or garlands for sorrow. We eat them, make our furniture from them, and the frames and doors of our homes.

opposite: *Salix caprea* – Pussy Willow (Salicaceae) – pollen grain, unexpanded condition, oblique polar view. [SEM x 5000]

overleaf: *Salix caprea* – Pussy Willow (Salicaceae) – close up of furry male catkin with anthers exerted.

page 94: *Alnus glutinosa* – Alder (Betulaceae) – close up of male catkins, female flowers above.

page 95: *Alnus glutinosa* – Alder (Betulaceae) – close up of female flowers.

page 96: *Alnus glutinosa* – Alder (Betulaceae) – catkins dispersing their pollen; note the old (last year's) female cones above.

page 97: *Corylus avellana* – Hazel (Corylaceae) – pollen grain, not fully expanded. [CPD/SEM x 2000]

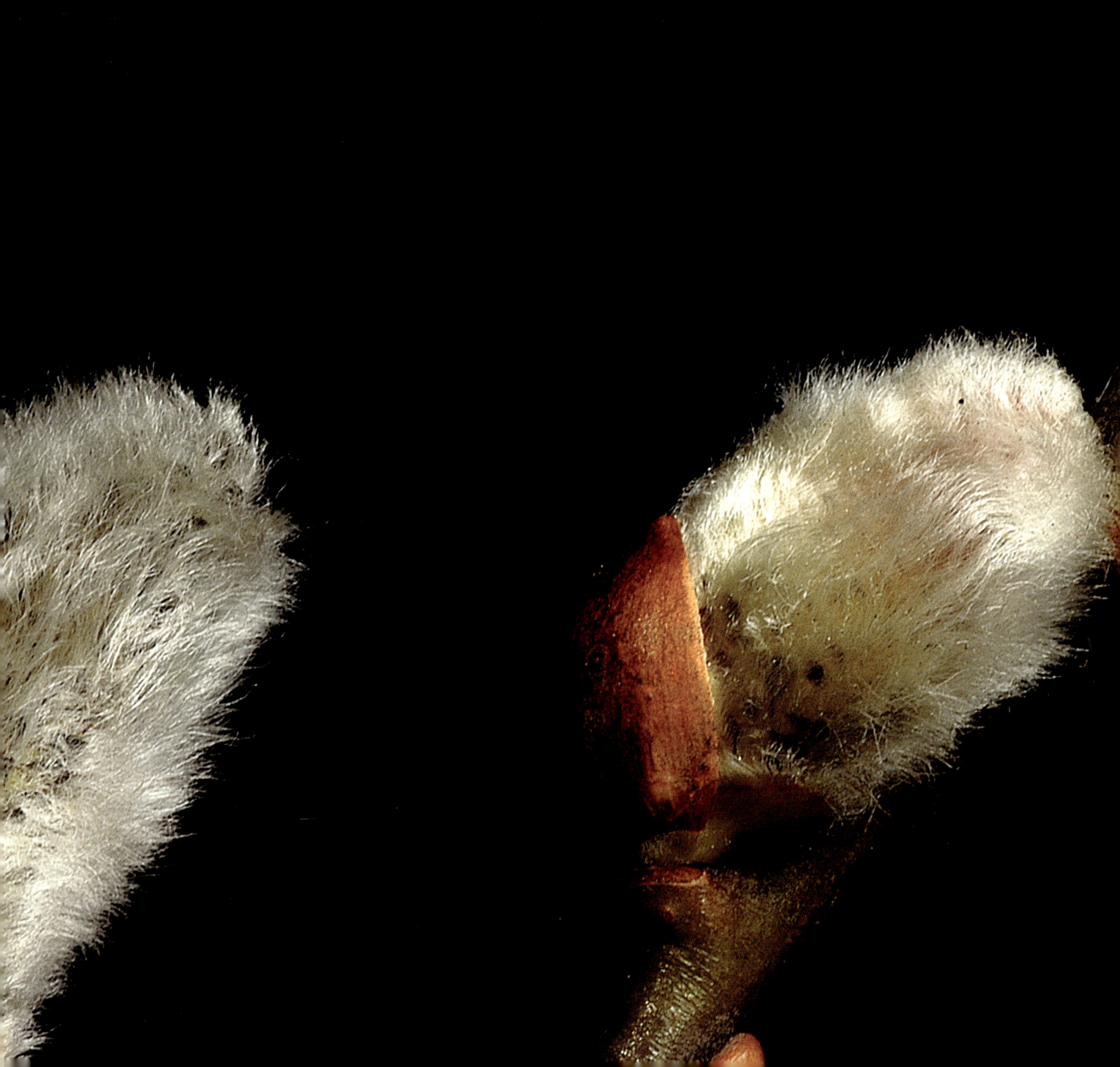

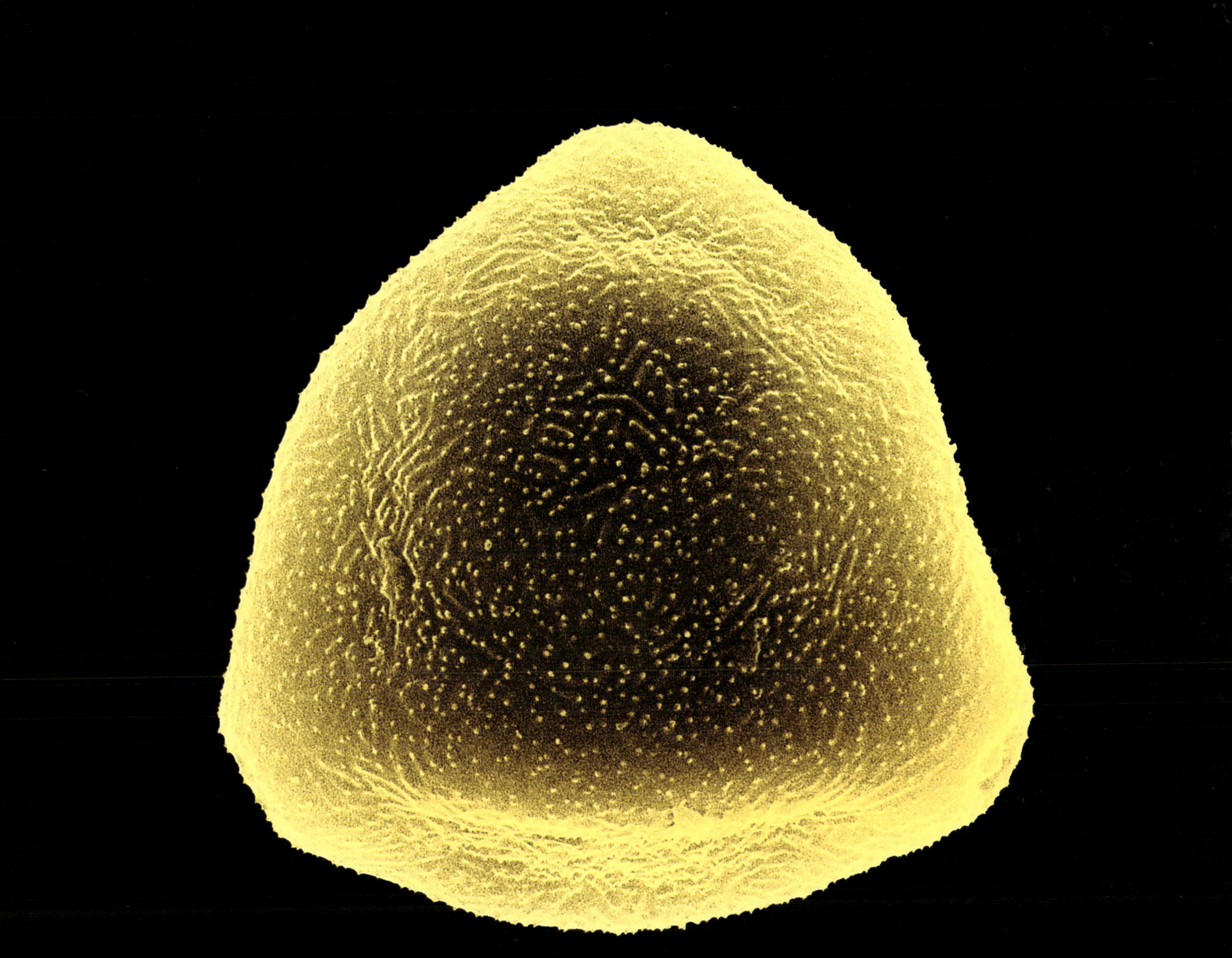

They may be rapidly reproducing weeds such as Groundsel, Stinging Nettles or Dandelions; or, like Daffodils, Tulips and onions they may develop from bulbs. We may eat plants before they ever flower, and fail to think of them as flowering plants – cabbages, celery, lettuces, and so forth. Or the flowers get rather ignored because we are waiting for the fruit (squashes and tomatoes), seeds (peas), tubers (potatoes) or roots (carrots and radishes). We may grow them for their ornamental leaves and not even notice that they also have small, or insignificant flowers (*Coleus*, and the Prayer Plant – *Maranta*). Palms, grasses, duckweeds, cacti and succulents are flowering plants. All the prized hardwood trees such as Mahogany, Rosewood, and Sandalwood have flowers, as do our beloved Oaks and Beeches. But while all flowering plants produce pollen, it is not essential for reproduction in all plants.

ASEXUAL REPRODUCTION

Most plants and animals require cross-fertilisation for the production of healthy offspring. However, the strategy for achieving this is different in each of the two groups. Animals are either male or female, the two sexes do not co-habit in the same body. Of necessity each sex has to actively seek out the other sex in order to mate. For plants sex is much more passive; the two sexes frequently inhabit the same body, frequently the same flower, but successful mating is usually achieved via an intermediary, the pollinator.

A significant difference between most animals and plants is that animals must reproduce sexually, while plants often have a choice. The Strawberry is an obvious example, with its runners enthusiastically throwing up new plants; or the Spider Plant (*Chlorophytum*) with young plants developing on the flowering stems and, of course, many of us have had our successes with new individuals from leaf or stem cuttings, or by simply dividing up a plant clump at its root base in the autumn to make more plants. For higher animals this is the stuff of science fiction.

Primula cv. – Polyanthus (Primulaceae) – section through a 'pin flower' showing the arrangement of the ovules on the broad central placenta.

Primula cv. – Polyanthus (Primulaceae) –
section through a 'pin flower' showing stigma near
top of flower tube, with the anthers much lower down;
note the sectioned ovary with the young ovules visible.
Polyanthus garden cultivar.

opposite: *Primula* cv. – Polyanthus (Primulaceae) –
'pin flower' – stigma exerted first.

Primula cv. – Polyanthus (Primulaceae) – section through a
'thrum flower' showing stamens near top of flower tube,
with the stigma much lower down.

opposite: *Primula* cv. – Polyanthus (Primulaceae) –
'thrum flower' – stamens exerted first.

Why do plants have a choice, but animals do not? At first glance this seems unfair. However, looked at another way, why do animals have legs, and sometimes wings as well, to escape from predators, but plants do not? This too seems unfair.

It is all to do with survival of the species. Asexual reproduction in plants can be viewed as a neat back-up strategy when sexual reproduction fails. However, this is cloning – the plant produces a copy of itself. Plant breeders clone to maintain the characteristics of their specially bred varieties; they want to prevent the variety from 'reverting', which is the likely outcome if the plants are allowed to cross-pollinate. In some plants, such as weedy annuals, asexual reproduction is often poorly developed. However, in other plants, such as Strawberries and their relatives, it is a highly successful alternative. Nevertheless, without sex very few plants produce fertile seeds.

Sexual reproduction is a much more successful strategy than asexual reproduction – cross-fertilisation is essential for maintaining the vigour of the species, whether animal or plant. Think, for example, of a remote community or village where everyone is inter-related. Not only does the IQ level of the individuals tend to suffer as a result, but any inherited physical defects stand more chance of being passed around the group than of being eliminated by 'new blood'. In time not only does the health of such communities become vulnerable to shared weaknesses, but they may also succumb to invasion by more vigorous and rapidly multiplying cross-bred outsiders.

In spite of the possibility of asexual reproduction, and a plant not having the freedom to move from place to place, evolution has, for reasons already outlined, strongly favoured sexual reproduction. It is what most plants do, most of the time, and they have a remarkable number of strategies for effecting cross-fertilisation. Nevertheless, although the majority of flowering plants have evolved to be self-infertile ('self-incompatible'), many species are able to self-fertilise if there are no pollinators to carry the pollen, from the anthers of one

Pinus tabuliformis – Chinese Red Pine (Pinaceae) – pollen being released on the wind from the male cones.

plant, to the receptive stigma of another plant of the same species. A minority of plants are habitual self-fertilisers.

HOW TO AVOID SELF-FERTILISATION

One way that plants avoid self-fertilisation is by being dioecious (from the Greek, meaning 'living in separate houses') with separate male and female plants. Most gardeners at some time or another have tried to ensure that they have both a male and a female plant of a species with showy berries such as hollies or *Skimmias*. To be sure of getting this right it is important to look at the plants when they are flowering. Take a hand lens or magnifying glass to the nursery or garden centre, and look to see if the flowers have well-developed anthers and no pistil (or a very rudimentary one), or a well-developed pistil and no anthers (or very rudimentary ones); then buy one of each type.

Alternatively, a species may be monoecious with separate male and female flowers on the same plant (from the Greek, meaning 'living in the same house'). Hazel (*Corylus avellana*) is a good example, with showy male flowers (catkins) and tiny, red, star-like female flowers in the early spring.

Interestingly, from the point of view of promoting sexual vigour and avoiding self-fertilisation, about eighty per cent of flowering plants are hermaphrodites, that is, they have both stamens and a pistil present in the same flower.

Clearly some careful strategic planning is necessary to achieve cross-fertilisation, and hermaphrodite flowers have developed a number of different strategies:

Self-incompatibility

There are two types of self-incompatibility. In gametophytic plants the female organ does not recognise, and blocks the passage of, the germinating pollen tube carrying the sperm (male gamete) into the stylar canal – this is a common

Dactylorhiza fuchsii – Common Spotted Orchid (Orchidaceae) – being visited by a species of Longhorn Beetle (suborder Phytophaga, family Cerambycidae).

syndrome and is considered to have developed very early in the evolutionary history of flowering plants.

In the less common sporophytic mode of self-incompatibility the surface of the stigma does not recognise the chemical signals given off by the external coatings of the pollen grain. This either inhibits the trigger for pollen tube growth to commence, or the pollen tube is rapidly rejected as it tries to penetrate through the stigma into the stylar canal.

Heterostyly

Varying the length and/or position of the male and female parts of the flower is an effective but uncommon strategy. The best known British example is the Primrose (*Primula vulgaris*); it can also be seen very clearly in Polyanthus flowers (a garden cultivar of mixed ancestry which includes the primrose and the cowslip). In the Primrose two arrangements are recognised – 'pin' flowers with a long pistil and stamens halfway down the flower tube, or 'thrum' flowers with the stamens at the top of the flower tube and the tip of the pistil reaching only half way up.

Heteromorphy

A variant of heterostyly occurs in Purple Loostrife (*Lythrum salicaria*) based simply on varying the length of the pistil: short, medium or long, but only rarely at the same height as either of the whorls of stamens.

Dichogamy

This is a very common strategy. By varying the timing of maturity between the stamens and the pistils within a flower, the sexually active pollen has to be transferred to another flower with a sexually active stigma. More frequently, within a flower, the stamens mature first ('protandry') but in some flowers the pistils are first to mature ('protogyny').

Echium vulgare – Viper's Bugloss (Boraginaceae) – being visited by a Bumblebee (female *Bombus terrestris* L.).

Insects

With the exception of self-pollination ('autogamy') discussed earlier, the only pollination syndrome not requiring a third party (vector), flowers have a wide range of pollinators. Insects are the most ancient of categories, they are also the largest group of pollinators. The most important are social insects, especially bees, butterflies, moths and beetles. Many have evolutionarily specialised plant relationships. Bees are very efficient pollinators, and many plants have co-adapted with bees to their mutual benefit, but not just honey bees (discussed below). The honey bee is just one species of a very large group which includes, not only hundreds of species of bees but also, the wasps and the ants. However, while bees actively collect both 'nectar' and pollen to sustain the bee colony, ants pollinate by accident, rather than design, as they crawl in and out of blossoms in search of sugar (nectar) or pollen as food. Mutually beneficial ant–flower adaptations are uncommon but an unusual, highly specialised 'symbiotic' wasp-flower relationship is described further on. Fossil records of social and solitary bees do not seem to extend further back than about 85 million years ago. Nevertheless, other lineages of the group which includes the bees (Hymenoptera) notably the Saw Flies (Symphyta) are probably as old, or older, than the beetles (more than 230 million years).

Of the other important insect pollinators, butterflies, and moths, both have a long tongue, or correctly 'proboscis', a specially adapted feeding/sucking food canal, which curls up, like a spring coil, under the head of the insect, when not in use. Butterflies and moths are highly adapted to dip for nectar in tubular flowers, their main source of food. Adaptation in moths goes even further, many night-flowering species are heavily scented to attract moths; these odours are, incidentally, often attractive to humans as well. It is interesting to note that the earliest fossil records of moths, which precede the butterflies, occur at about the same time as the earliest evidence of flowering plants.

Lotus corniculatus – Birdsfoot-trefoil (Leguminosae) – being visited by a Six-spot Burnet moth (*Zygaena filipendulae* L.).

Pollen: The Hidden Sexuality of Flowers

"Now for the hive itself. Remember, whether you make it
By stitching concave bark or weaving tough withies together,
To give it a narrow doorway: for winter grips and freezes
The honey, and summer's melting heat runs it off to waste.
Either extreme is feared by the bees. It is not for fun
That they're so keen on caulking with wax the draughty chinks
In their roof, and stuff to the rim of their hive with flowery pollen,
Storing up for this very job a glue ['propolis'] *they have gathered*
Stickier than bird lime or pitch from Anatolia."

[Virgil – The Georgics, Book IV.
Translated by Cecil Day Lewis]

Apis mellifera (Honey Bee) – a worker bee with full pollen baskets (corbiculae).

Flies and beetles are often rather generalised in their approach to flower visiting and, like the ants, tend to pollinate by chance as they visit flowers to forage, some beetles are actually quite destructive to the flowers they visit. Beetles are considered to be a very ancient category of flower visitors; fossil records of beetles extend back further than the first flowering plants, by about 100 million years. From primitive beginnings, at least 190 million years ago, some lineages of flies have become more highly evolved flower-visitors. In some species the short fly-like proboscis, typical of many flies, has evolved into a much longer organ, adapted to visiting more evolved flower forms. There are flowers which have evolved odours to attract carrion and dung flies, notably some species of Araceae (Arum Lily family), Asclepiadaceae (Milkweed family) and Orchidaceae (Orchid family). Unlike moth-pollinated flowers, the odours of these carrion and dung fly attracting blossoms are usually very unpleasant to the human nose.

The honey bee

Honey bees, *Apis mellifera* L., are social insects, and they have been an important part of animal husbandry in human civilisation for thousands of years – there are records from Egypt from as early as 3000 BC, where honey bees, hives and beekeeping practice are often depicted on early stone bas-reliefs. Virgil, in about 30 BC, writes in 'The Georgics', "Next I come to the manna, the heavenly gift of honey." He then goes on to describe beekeeping practice in great detail.

Honey bees can only survive as part of the social colony. In summer a colony would comprise of between 50 – 70,000 bees. Within the colony the honey bees live on the 'combs' they create from body wax for the incubation of bee larvae, and for the storage of nutrients for the colony: honey (from nectar and honeydew) and pollen ('bee bread' for the larvae). There are three types of *Apis mellifera* in the colony: only one, the queen, is unique. The majority of the bees in the colony are undeveloped females (worker bees) they

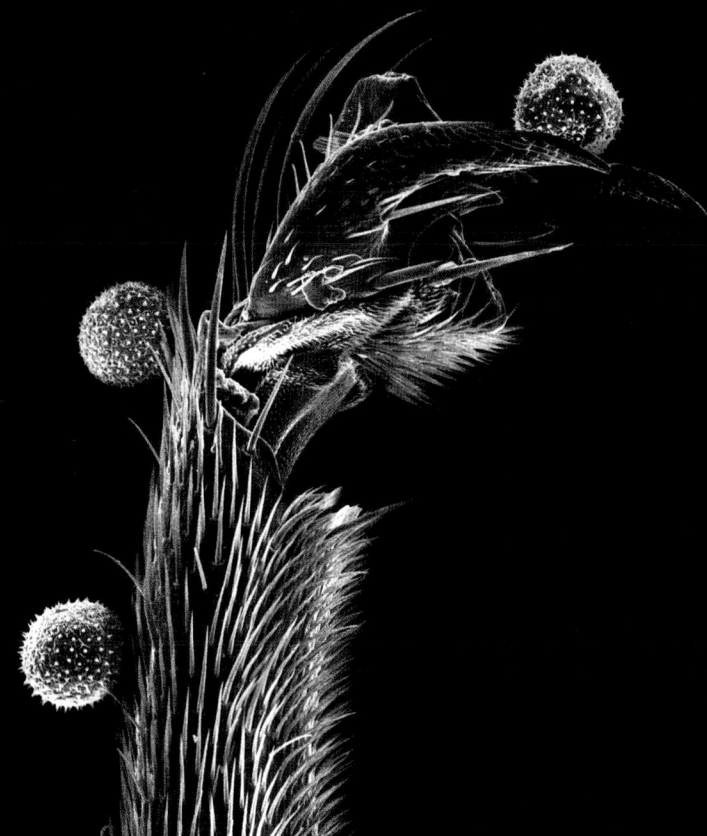

Pollen of *Malva sylvestris* – Common Mallow (Malvaceae) –
on the rear leg of a Bumblebee *(Bombus terrestris* L.).
[SEM x 100]

opposite: *Apis mellifera* (Honey Bee).

are rather smaller than the queen, while the largest bees, the drones (male bees), number around 2000. While the queen and worker bees are the result of fertilised eggs, the drones develop from an unfertilised eggs ('parthenogenesis' or 'virgin birth'). Parthenogenesis can occur in plants or animals, although it is an unusual method of reproduction, because it results in cloned individuals. There are two type of parthenogenesis, diploid and haploid. Honey bee drones are the result of haploid parthenogenesis in which the eggs from which they develop are produced by meiosis in the usual way and are haploid. As a result the drones that develop from these unfertilised eggs also have haploid cells. The fertilised female eggs develop into adults with normal diploid cells. The queen is the only female with fully developed ovaries and, during the active season between March and October, she will lay between one and two thousand eggs per day. The only function of the drones is to mate with the young or 'virgin' queen; she will mate with about five to eight drones during her sexually receptive cycle of 5-6 weeks. Mating takes place on the wing, and during copulation the drone expires. Once the young queen has finished mating and started a good brood nest, the remaining drones (the majority) no longer serve any useful function. The worker bees expel them from the hive or kill them.

The worker bees have many tasks – but there is a strict division of labour. The average life span of a worker bee is about 40 days. For the first half of their lives they work within the hive, feeding the developing larvae ('nurse bees'), receiving nectar from the 'field bees', cleaning the hive, building the wax combs to house larvae and nectar, guarding the hive and preliminary 'orientation flights' around the hive. The queen larvae are fed on 'royal jelly' a food substance produced by nurse bees. The wax for the combs is excreted by the worker bees from special glands located on the underside of the body, between the abdominal segments. The wax cells include not only the usual sized hexagonal cells, but also a proportion of larger cells for the queen to lay the drone eggs, and a large asymmetric queen cell. After the hive period the worker bees become the field bees, collecting nectar, honeydew and pollen.

The sugary nectar solution is an attractant most commonly produced by special glands ('nectaries') in insect pollinated flowers. The worker bee absorbs nectar by penetrating her proboscis deep into the flower where the nectaries lie hidden – in the process her body becomes covered in pollen. She withdraws the nectar into her 'honey sac' (stomach), absorbing only as much as she needs for survival. During her return to the hive, water is being drawn away from the nectar in the honey sac and passed into the intestines. On return to the hive the contents of the honey sac are unloaded into empty cells in the comb via her proboscis or, she passes the nectar to the hive bees to process and store in empty cells. The elimination of water from the nectar, and the addition of enzymes produced by the worker bees are essential processes in converting nectar to honey, and this task is continued by the hive bees. Some of the bees are positioned between the comb and the hive entrance, fanning their wings very fast to shift the water evaporating from the honey out of the hive. When the water content has been reduced to less than 20% the honey is ready. The full honey cells are sealed over with a thin layer of wax.

Honeydew is produced by certain species of aphids that suck sap from different species of trees, to absorb nutrients. The waste sap is excreted as honeydew and remains as shiny droplets on pine needles or leaves of other trees. It can be a nuisance, for example, in the UK in summer, honeydew from Lime trees – *Tilia x vulgaris* – a commonly planted street tree, frequently coats cars parked underneath. Bee colonies located near plantations of notably, conifers such as pine, fir and spruce, harvest the honeydew from mid summer. Honeydew from conifers is frequently referred to as 'forest honey', while that from 'deciduous' trees is called 'leaf honey'. The honey bees process honeydew in the hive in the same way as honey. However, it is considered to have an inferior flavour. Commercially it has to be distinguished from nectar honey.

While nectar and honeydew are collected for carbohydrates, pollen is

Pollen loads from the corbiculae of *Apis mellifera* (Honey Bee) – the differences in colour represent some of the spectrum of pollen colour from different species of flowers foraged by the worker bees.

collected as a source of protein. As the bees fly from flower to flower in pursuit of nectar, they remove the pollen picked up on their body hair by using the brush–like hairs on their legs. In a related action the pollen is collected into the pollen baskets ('corbiculae') on their hind legs. They frequently use nectar collected in the honey sac to help moisten and stick the pollen grains together into 'pollen loads'. These loads are carried back to the hive where they are pulled off and pushed into empty cells. The young hive bees then compress the pollen into the cells with their heads and forelegs to make bee bread. It is possible to gain an idea of what bees have been foraging by examining the colours of the pollen loads. This is because bees collect systematically, targeting plant species in full flower, with abundant rewards. Pollen loads are usually representative of single species, although mixed loads also occur, where a pollen source has run out and another species is visited on the same collecting trip. Bees may be primarily on a nectar collecting trip and pollen collection is an incidental activity. They may also target flowers with no nectar, pollen being the primary objective.

A healthy bee colony including bees hatched from August onwards, plus a queen, will 'winter over'. They form a tight cluster, distributed over about four to eight combs. The cluster moves very slowly along the combs, using the pollen and honey stored in the combs to generate a heat of about 20-25° centigrade in the centre of the cluster. The external temperature of the cluster must not fall below 7-10° centigrade. When outside temperatures rise slightly above 10° centigrade the bees will take 'cleansing flights'. This is because bees avoid defecating in the hive or colony. This is not a problem during the summer but, in winter, waste collects in their intestines and, until it is warm enough for them to leave the colony, they cannot excrete the accumulated waste.

Are honey bees the only insects that make honey for food?

There are some tropical paper wasps (the genus *Brachygastra*) that make small amounts of honey and store it; and also the genus *Polybia* in South America. A

Apis mellifera (Honey Bee) – a worker bee foraging for pollen and nectar on *Ceanothus* cv. (Rhamnaceae); note full pollen baskets (corbiculae).

few species of ants store honey in the bodies of the workers, other than that all the honey-making insects are restricted to one small lineage of bees (orchid bees, bumblebees, stingless honey bees and honey bees – *Apis mellifera* and a few other species of *Apis*). Well over 12000 of the 16000 known species of bees do not make honey. Only insects that live in colonies are associated with honey production. This is because honey is a long-term resource, lasting longer than the lifespan of an individual insect, but not longer than the life of the colony. About 150 or more species of Bumble bees make a kind of honey to sustain themselves during brief spells of bad weather. In Central America there are many species of stingless bees that make honey, for example from the genera *Melipona Mourella*, *Plebeia* and *Trigona*. The production of honey from these bees is much less than from *Apis mellifera* but, nevertheless, it is traditionally harvested and enjoyed by local people in a number of Central and South America countries. The formal definition of honey by the European Codex Alimentarius states that it is produced only by *Apis mellifera*. However, the 'honey' produced by bees other than *Apis mellifera* is reported to be sweet and delicious and looks very much like honey.

Animals

Other animal pollinators include bats and other small mammals or marsupials, birds and even snails. Bats are highly evolved and, like moths, are nocturnal. Bats are mostly insectivorous and have a highly developed sonar system. However, species of the tropical New World subfamily Glossophaginae (in suborder Microchiroptera – which includes the insect-eating bats) have evolved as fruit-eaters, while in the Old World another suborder, Megachiroptera, feed almost exclusively on nectar and pollen. In these bats the sense of smell is highly developed, while their sonar system is less well-developed than in the insect feeders. Among the bird pollinators are the short-billed Passerines, the wonderfully adapted long-billed Humming Birds of the Americas and the Australian Honey-eaters. Small vegetarian mammals,

Ophrys sphegodes – Early Spider Orchid (Orchidaceae); note the two pollinarium near the central part of the flower, just above the swollen labellum.

opposite: *Ophrys sphegodes* – Early Spider Orchid (Orchidaceae) – chalk downland, Kent.

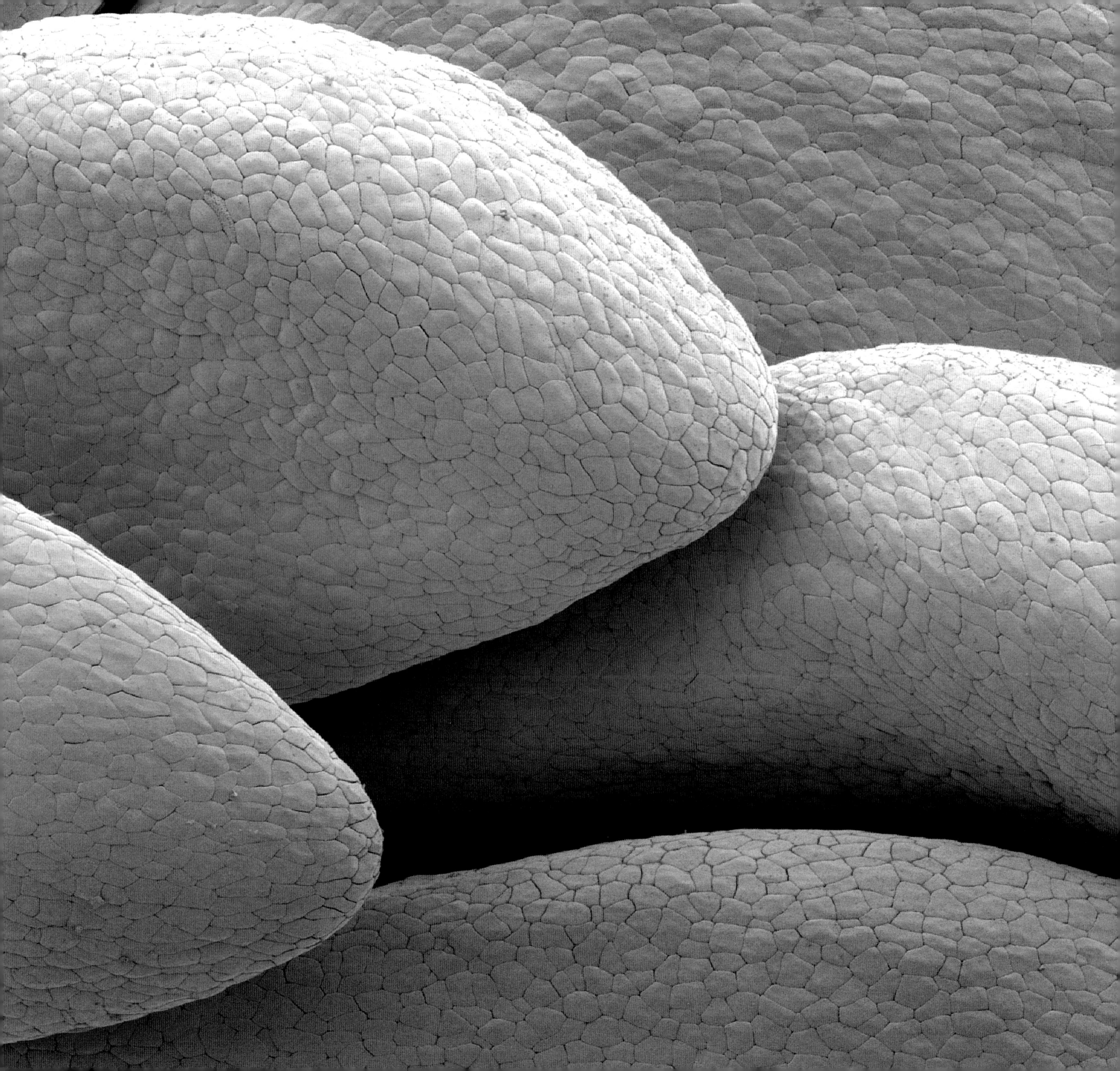

opposite: *Calanthe aristulifera* (Orchidaceae) – close up of pollinium [CPD/SEM x 100]

below: *Calanthe aristulifera* (Orchidaceae) – complete pollinarium: from the base of the image are the pollinia which are attached to a branched caudicle and, at the base of the caudicle there is a viscidium, this attaches to an insect visitor to allow the pollinarium to be transported to another flower for fertilisation. [air dried/SEM x 25]

overleaf left: *Ophrys sphegodes* – Early Spider Orchid (Orchidaceae) – close up of the pollinarium. [CPD/SEM x 650]

overleaf right: *Ophrys sphegodes* – Early Spider Orchid (Orchidaceae) – close up of the pollinarium. [CPD/SEM x 150]

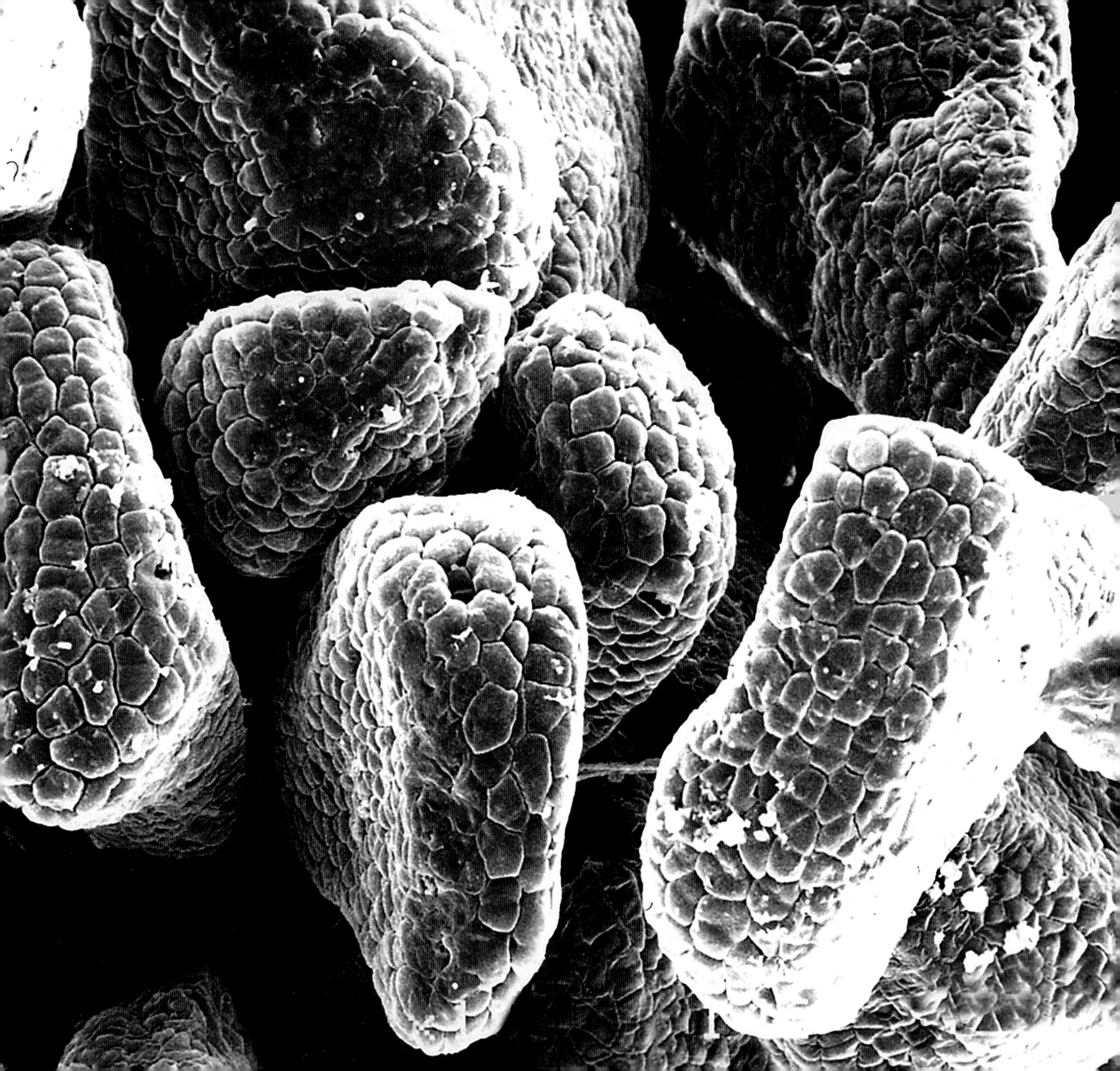

Lamium orvala – (Lamiaceae) – flower in profile, an example of zygomorphy.

opposite: *Lamium orvala* – (Lamiaceae) – flower seen from front, another example of zygomorphy.

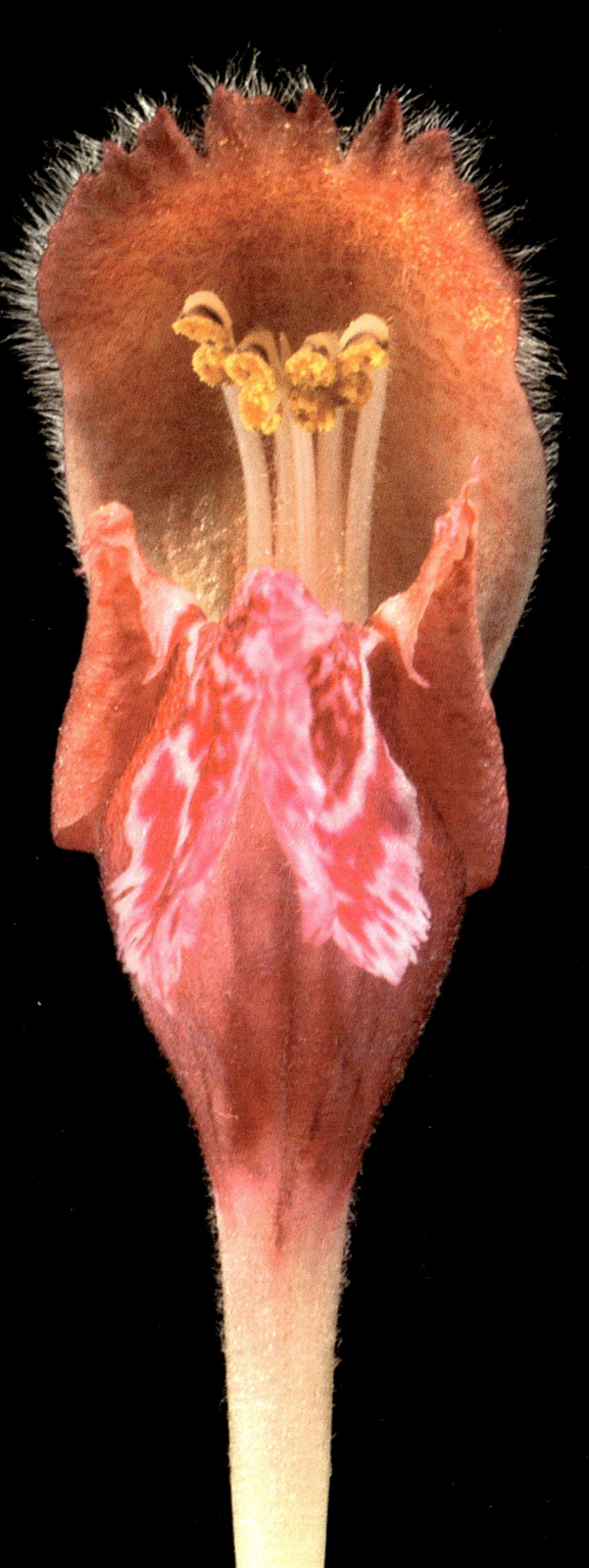

especially in the tropics and sub-tropics, often transfer pollen as a result of foraging activity, for example in Hawaii small nocturnal rats – White Eyes – are reported to climb around *Freycinetia arborea* trees to nibble the juicy bracts. Bats are also attracted by the fleshy bracts of *Freycinetia* species, while in Australia there are a number of reports of small marsupials foraging and transferring pollen. Some show no signs of adaptation as pollinators, while others, such the Honey-mouse, are apparently quite highly adapted as pollinators. The Honey-mouse, which visits the long narrow flowers of the *Protea* family to feed on the nectar, has a strongly projecting snout, very reduced or absent teeth, and a very long narrow tongue with a brush-like tip. Snails are reported to be the pollinators of *Aspidistra* – an East Asian relative of our north temperate species *Convallaria majalis* (Lily of the Valley). The case of *Aspidistra* is curious. *Aspidistra* was popular as a Victorian parlour plant, beloved for its large, dark green strap-like clumps of leaves, and extraordinary tolerance of minimal watering, dust and dimly-lit drawing rooms, but never for its flowers. This is not surprising as the flowers often come and go unnoticed. They are small, chubby, insignificant, pale mauvy-brown and stemless, flowering at ground level. In the wild, snails supposedly crawl across the flowers, thus transferring pollen from one blossom to another. There are no other flowers recorded with snail pollination.

Wind and Water

Many plants are adapted to wind pollination including plants with tassel-like male inflorescences such as Alder, Oak, Birch and Hazel, the conifers and their relatives, grasses and nettles. The pollen grains of these plants tend to be small and/or dry, and produced in huge quantities, easily carried away on the wind. Grass inflorescences are finely developed for pollination by wind; the stems carrying the panicles of florets are usually long and slender, moving effectively even in a slight breeze. The individual hermaphrodite florets are without petals

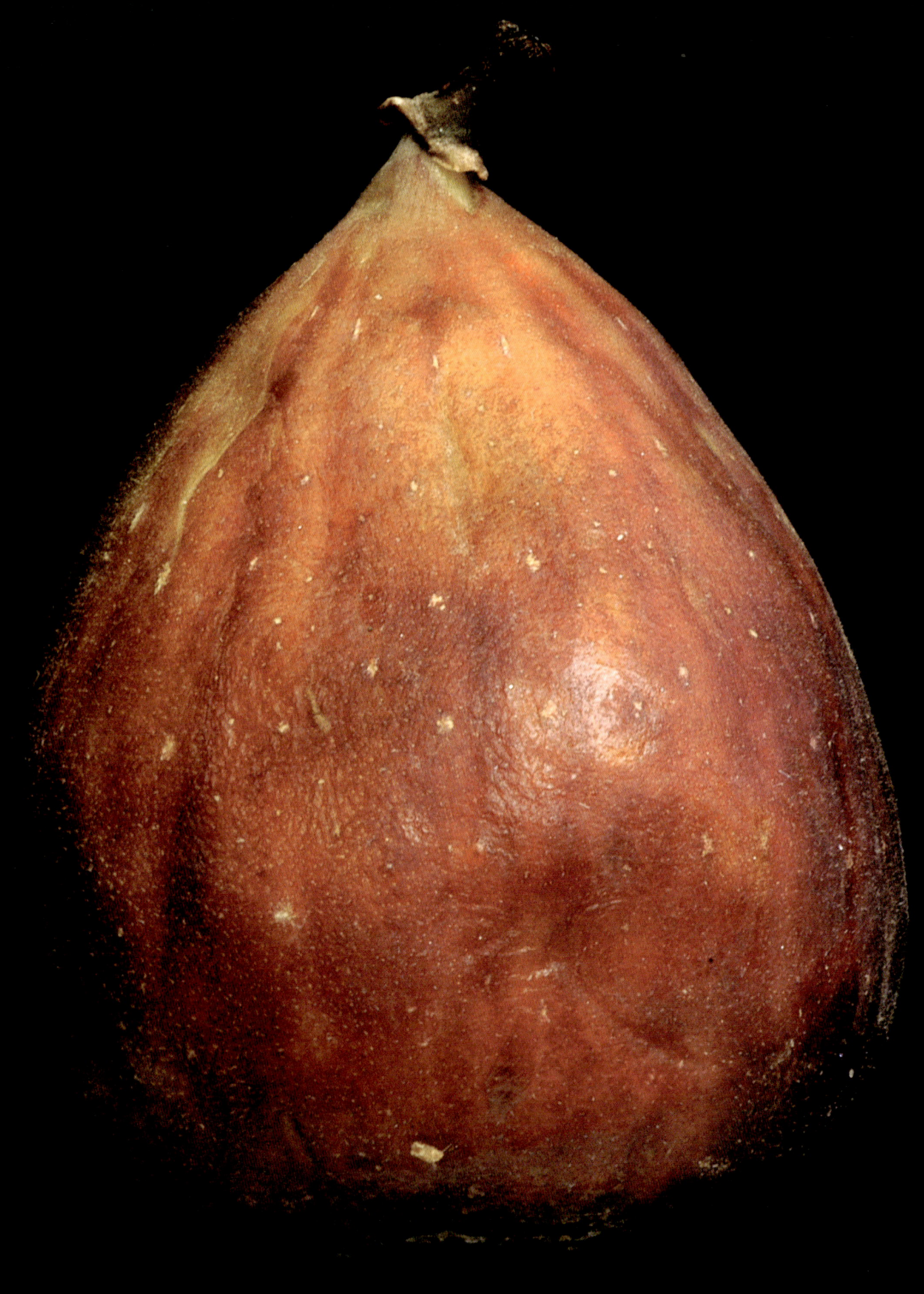

usually having instead two tiny scales ('lodicules'), two feathery stigmas to entrap the windborne pollen, and three stamens. The filaments of the stamens are thread-like with the anthers hanging loosely from their tips to catch the wind. Conifer pollen grains have a pair of air-inflated sacs. These sacci are evolutionarily associated with aerodynamics *and* hydrodynamics, because they allow the pollen to be successfully transported either by air or by water. Water pollination is well-developed in many freshwater dwelling plants such as Duckweed, as well as the Eel Grasses (*Zostera*), which have strangely adapted filamentous pollen grains about 2.5mm long. The grains are released in masses, and are passively carried by the tides across eel grass flats to curl around protruding female stigmas encountered en route.

CO-ADAPTATION

The earliest pollination vectors are now widely considered to have been insects, rather than the wind. Many insect groups had evolved before the evolution of the earliest flowering plants, including flies and beetles. Nevertheless, the extraordinary radiation and speciation of both insects and plants through the last 120-130 million years suggests that co-evolution of characteristics related to pollen transfer in plants has been highly beneficial to both groups. These include insect attractants such as nectar, the positioning of the nectar-producing organs ('nectaries') so that the insect has to brush past the pollen-producing organs (anthers) to reach the nectar, floral odours, ultra-violet colour guides and insect mimicry in plants; and the adaptation of mouth parts, such as the proboscis in moths, sense of smell, and ultra-violet colour vision in foraging insects.

There are some remarkable morphological co-adaptations between flower visitors (pollinators) and the flowers they visit, especially among zygomorphic flowers (i.e. flowers with bi-lateral symmetry). Such flowers are generally considered to be evolutionarily advanced. Well-known examples include

previous page left: *Ficus carica* – fig (Moraceae) – a greatly swollen receptacle enclosing numerous separate male and female (monecious) florets; note the peduncle (main flower stem) at the top of the fig.

previous page right: *Ficus carica* – fig (Moraceae) – section through a ripe fig to show the swollen receptacle (outer white area), enclosing the remains of the florets and swelling seeds in the pink fleshy centre. Note the orifice (ostiolum) at the base of inflorescence; this is where fig wasps would enter.

below: *Bellis perennis* L. – Common Daisy (Compositae)

opposite *Heracleum spondylium* – Hogweed (Umbelliferae) – a 'landing platform' inflorescence, comprising hundreds of tiny flowers; the petals of the outer flowers in each group are larger than those of the inner flowers.

overleaf left: *Heracleum spondylium* – Hogweed (Umbelliferae) – open locule of ripe anther with pollen spilling out. [SEM, naturally dried, x 100]

overleaf right: *Heracleum spondylium* – Hogweed (Umbelliferae) – a pollen grain, in dehydrated condition, lying on the surface of the anther. [SEM x 2000]

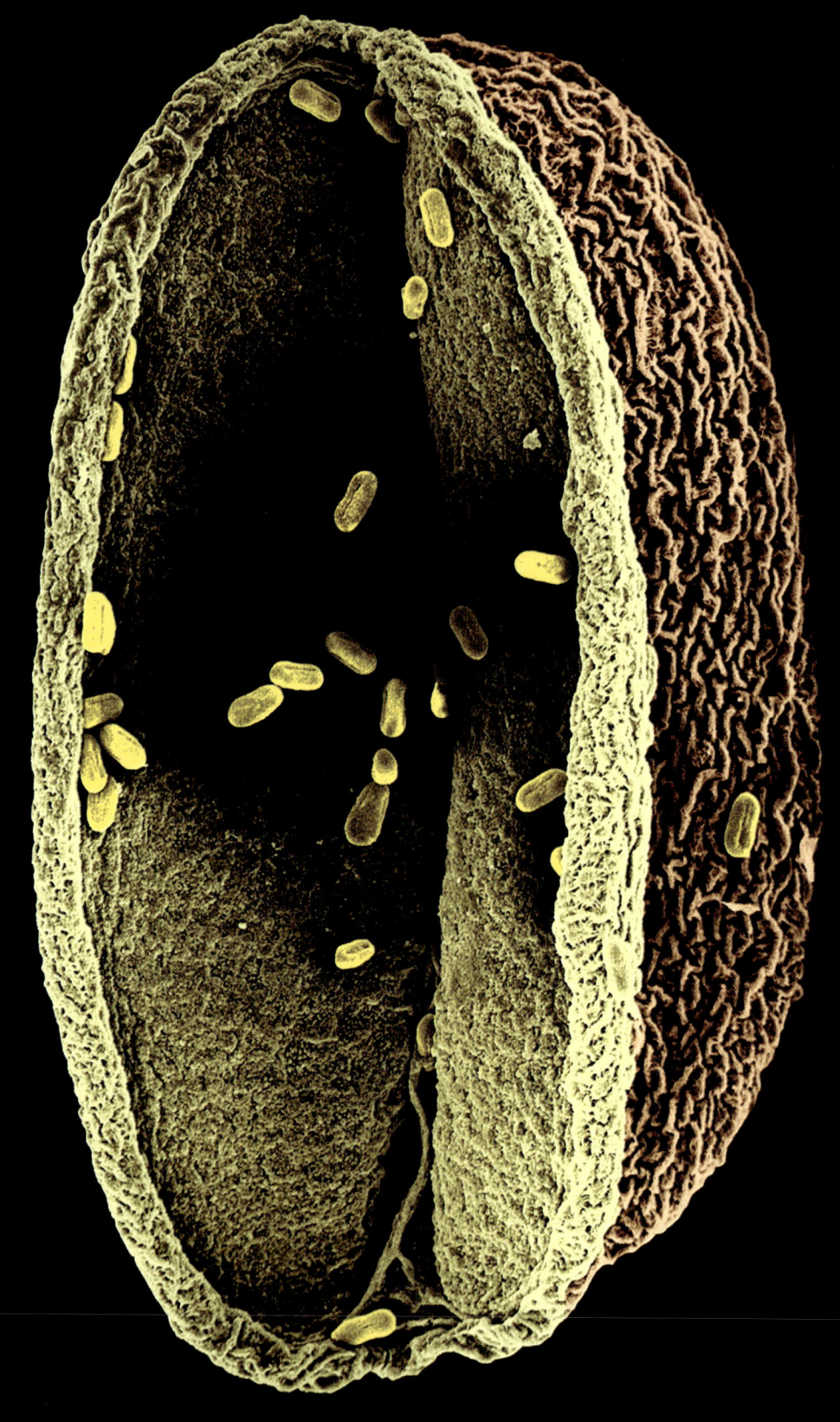

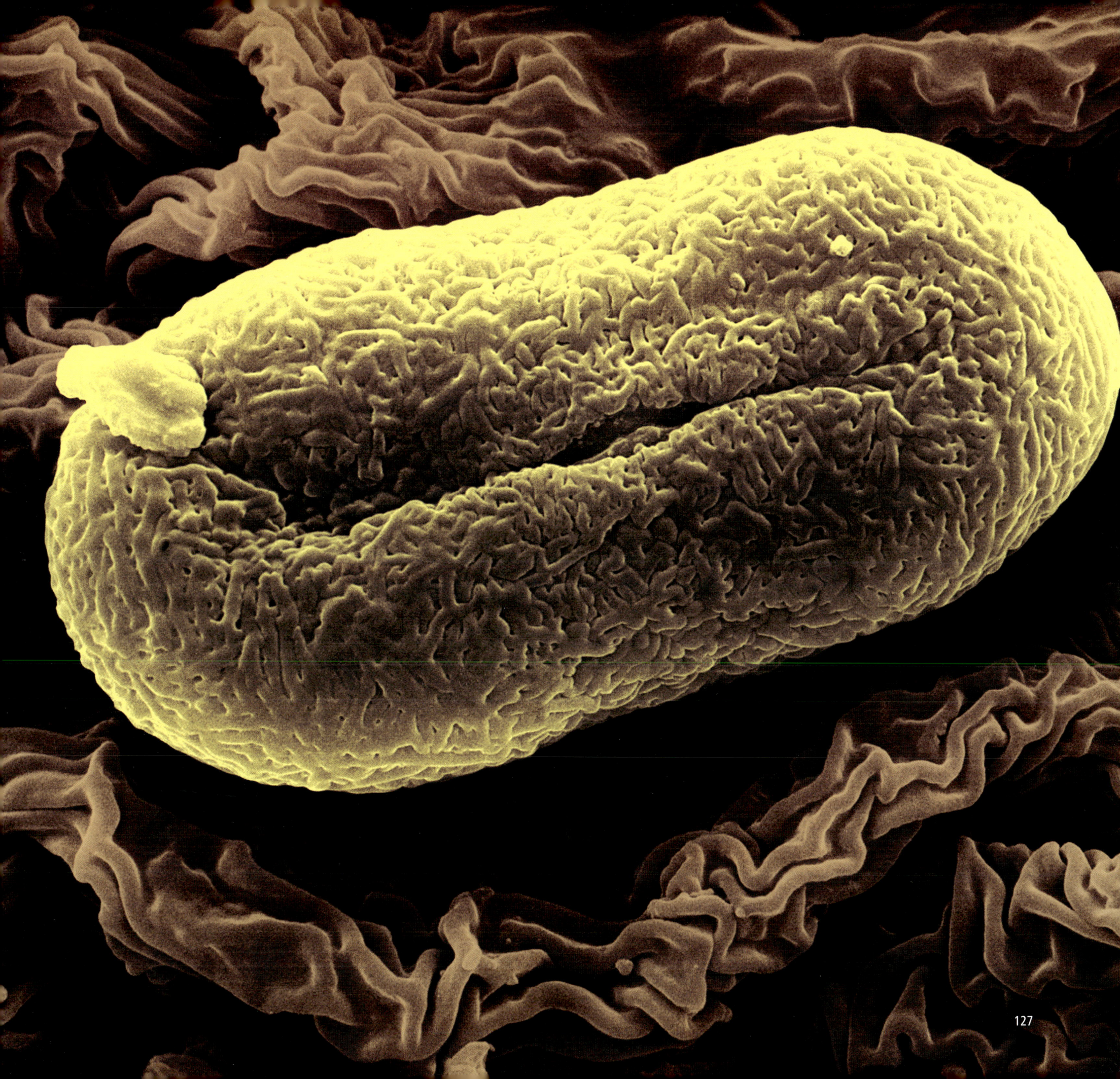

Zantedeschia cv. – Arum lily (Araceae) – close up of the spadix showing the junction between the male and female sections; note the pollen exuding from the anthers in strands; each strand, comprising many pollen grains, is stuck together by sticky pollenkitt.

opposite: *Gomphocarpus physocarpus* (Asclepiadaceae) – flowers dripping with nectar.

128

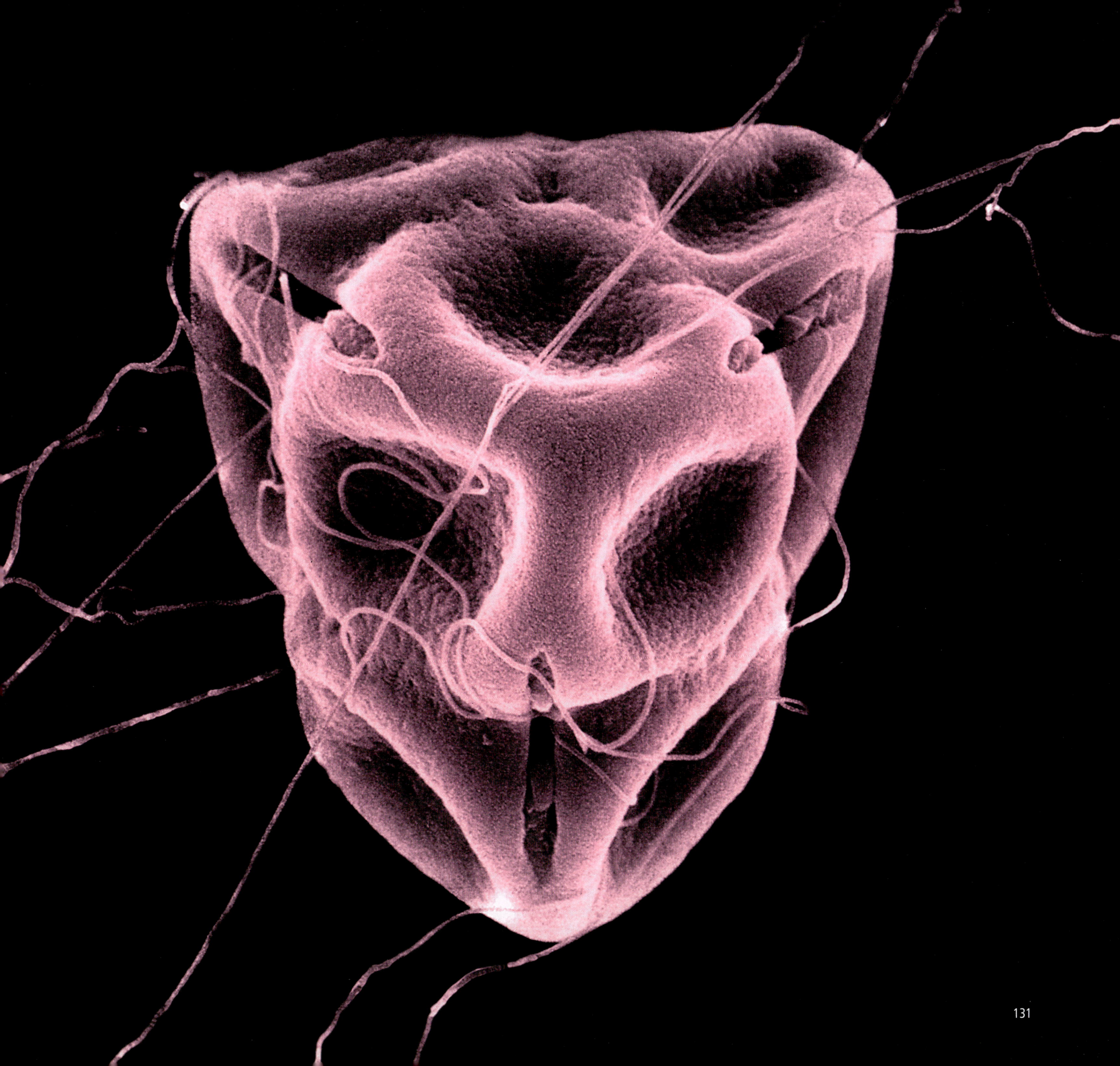

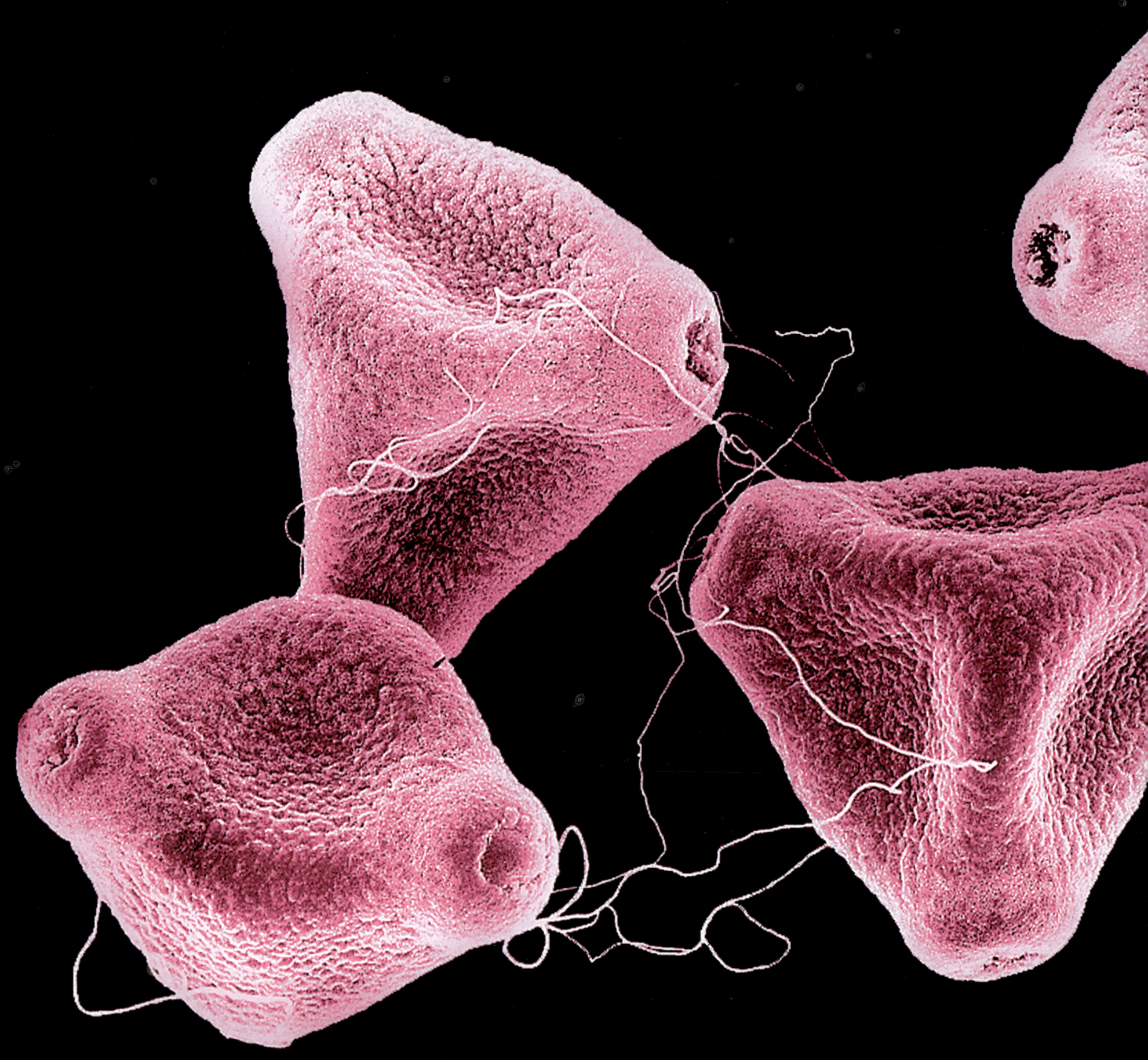

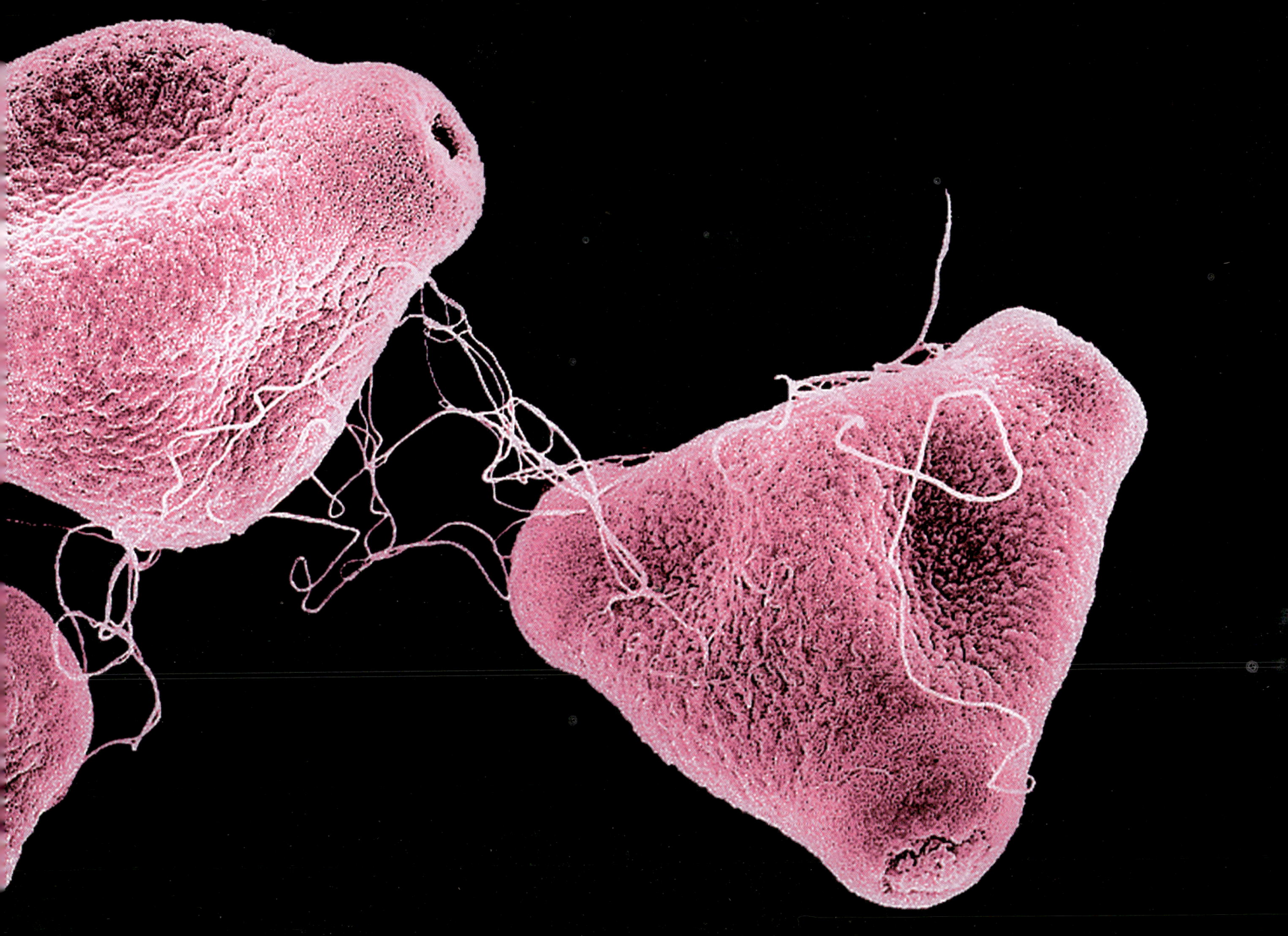

previous page left: *Rhododendron* cv. – 'Naomi Glow' (Ericaceae) – close up of flower; the filaments of the anthers have been loosely strung together by the numerous sticky viscin threads, which have linked the pollen grains in ropes as they left the anthers

previous page right: *Rhododendron* cv. – 'Naomi Glow' (Ericaceae) – a permanent pollen tetrad, typical of family Ericaceae, with numerous viscin tetrads, pollen grains in unexpanded condition. [SEM x 1000]

Epilobium angustifolium – Rose-bay Willow-herb (Onagraceae) – unexpanded pollen grains, loosely linked by long viscin threads. [SEM x 400]

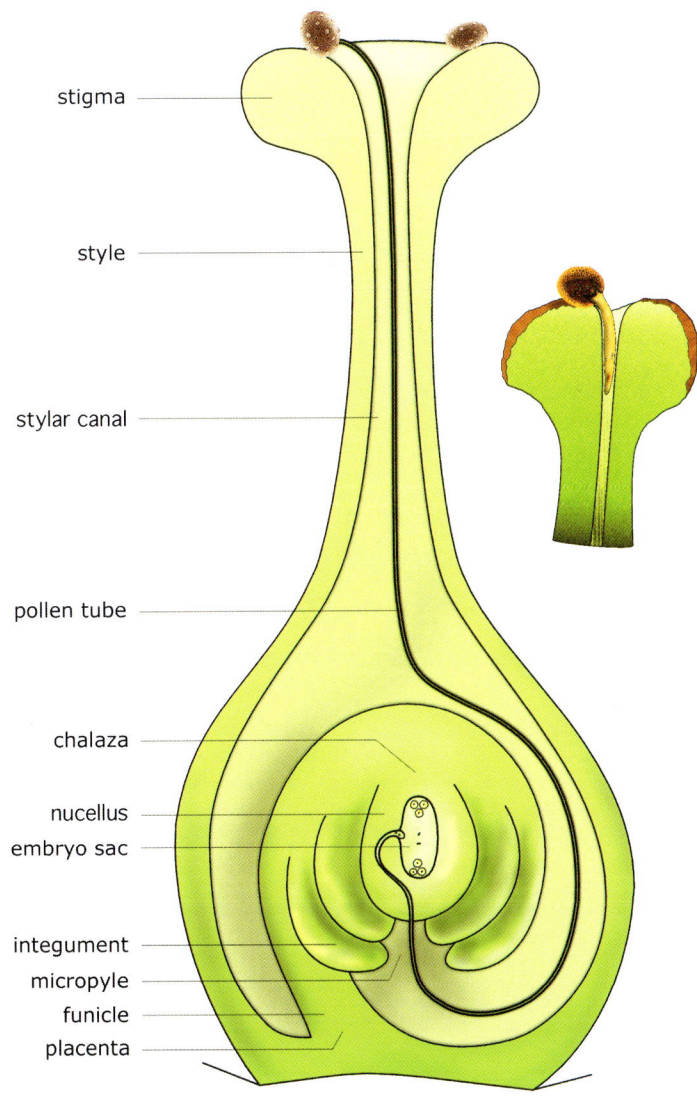

stigma

style

stylar canal

pollen tube

chalaza

nucellus

embryo sac

integument

micropyle

funicle

placenta

orchids that mimic species of flies, spiders or bees to persuade the insects to attempt copulation. Orchids belong to one of two large families where 'pollinaria', rather than anthers, contain the pollen. Pollinaria are very viscous; they are positioned near the top of the flower so that as the visiting insect re-emerges to fly away the pollinaria catch onto the head or body and are carried to another flower. Other zygomorphic flowers with very specialised insect-pollinator morphologies include Snapdragons, Monkey Flowers (*Mimulus*) and Foxgloves, as well as many members of the pea family (Leguminosae), and *Salvia* family (Lamiaceae).

The 'outside in' flowers, or rather inflorescences, of species of the Fig family (Moraceae) enjoy a mutually convenient relationship with different species of Fig Wasp. The inflorescences are very unusual in having a greatly swollen receptacle which encloses the tiny flowers – florets (comprising the inflorescence). This can be seen by cutting through a Fig: on the outside, the purple flesh is the swollen receptacle, the central soft, juicy, vivid pink, seedy part comprises the ripened florets 'gone to seed'. At flowering stage, when the wasps enter the top of the young Fig (the small, scab-like area opposite the stalk) they carry out two types of activity: pollination of the florets that have long-styled pistils, and ovi-deposition into the pistils of the short-styled florets. Subsequently the long-styled flowers give rise to seeds, and the short-styled florets develop into galls in which the next generation of Fig Wasps are incubated.

In some species the flowers, often very small, are closely associated in groups (inflorescences). Some inflorescences are arranged in a plate-like formation, and act as insect landing pads. Examples include the daisies (Compositae), where an outer circle of tiny florets, each with a well-developed petal, surrounds a circular pad of tightly packed florets without petals. Many members of the family Umbelliferae, such as Lady's Lace, Hogweed and Cow Parsley, as well as the Elder (Caprifoliaceae) have more loosely arranged

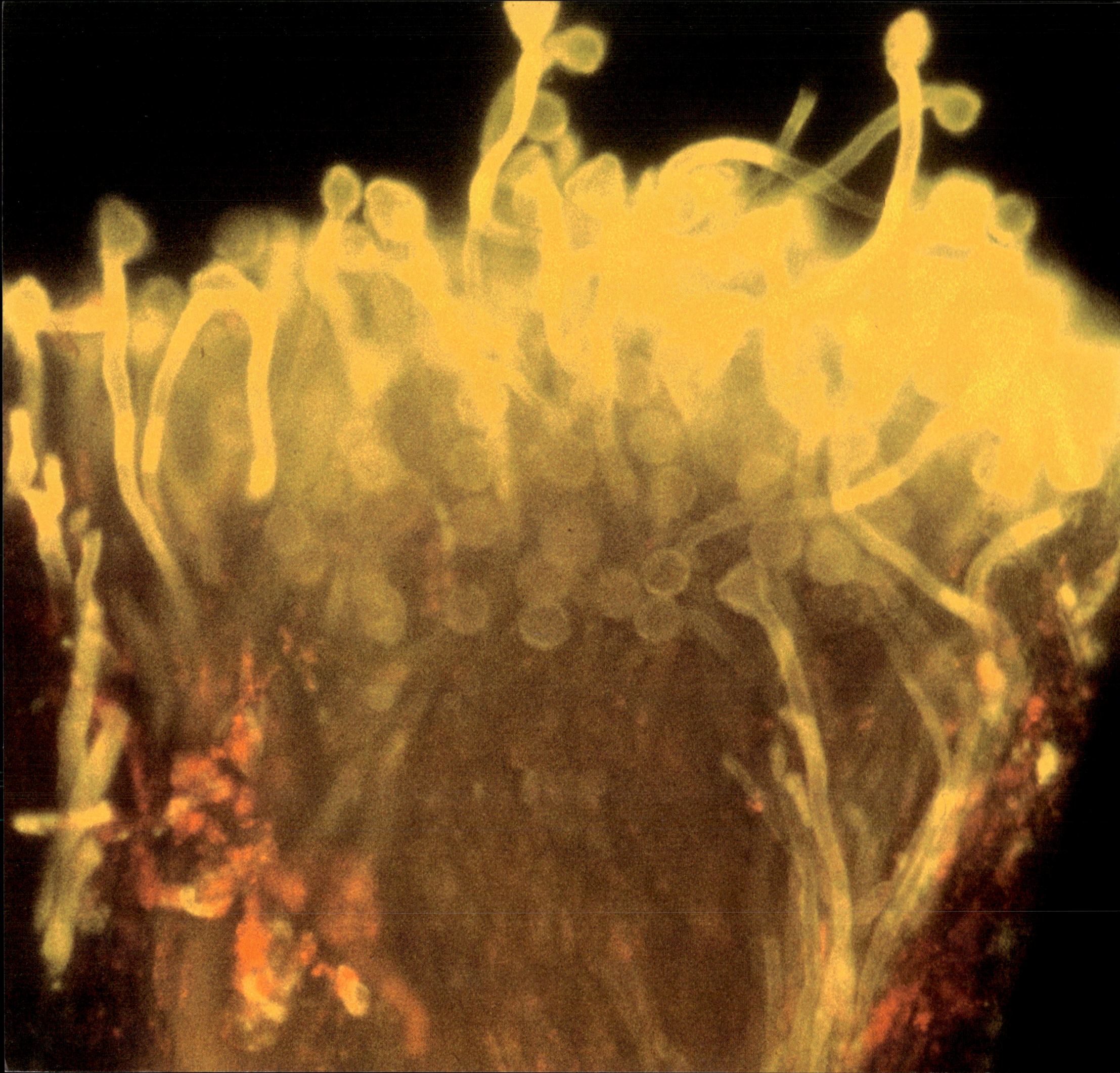

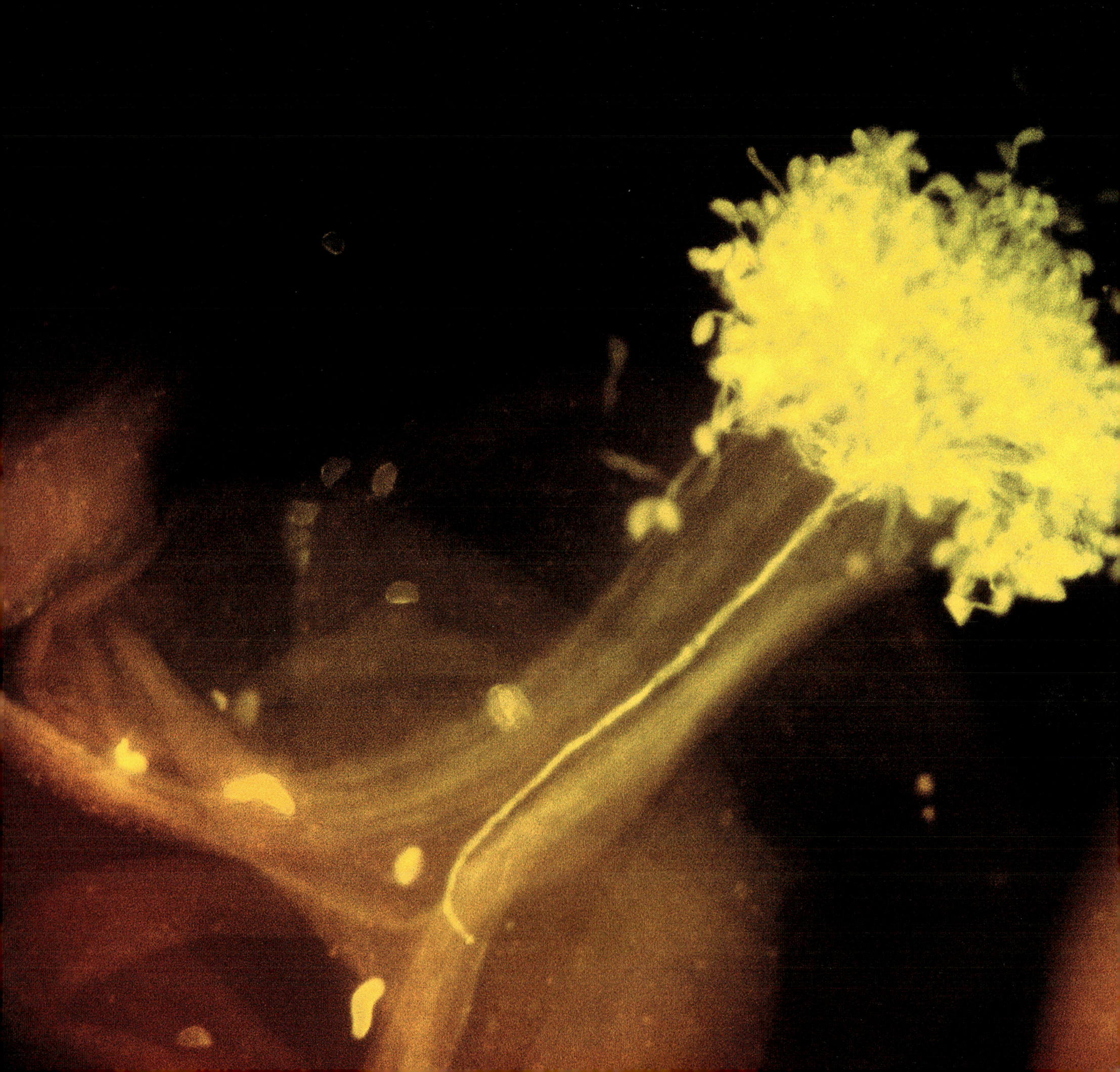

138

landing pad inflorescences which, in early summer, can be seen playing host to a cosmopolitan assembly of insect visitors.

Nectar is an important floral reward for flower visitors including, of course, bees. Humming Birds, Honey-eaters, as well as many moths and butterflies, have evolved either long beaks or long tongues to dip for nectar at the base of long corolla tubes. Many flowers have nectar guide marks on their petals. Some of these marks are visible to the human eye, but more often they are only visible in ultra-violet light, to which insect vision is adapted. To insects nectar guide marks are very clear.

Floral scent is another inducement – flowers, heavily scented during the evening or night, attract moths and nectar or fruit eating bats. Evil-smelling flowers, such as some of the Arum Lilies (Araceae), to which carrion and dung flies are attracted, only to be enticed and entrapped into the swollen base of the enlarged 'spathe' where the male and female flowers encircle the base of the 'spadix'.

POLLEN AS A POLLINATOR ATTRACTANT

In some groups of plants such as the *Rhododendron* and Heather family (Ericaceae) and Evening Primrose and *Fuchsia* family (Onagraceae) the pollen grains have masses of tiny thread-like structures ('viscin threads') attached. The threads not only cause the pollen grains to clump together, but also act as devices for attaching the grains to insects. The external oily lipid component of pollen grains helps stick them to the body parts of flower visitors. Furthermore, the lipids have distinctive odours that belong to the same chemical classes of aromatics as those found in flowers.

However, there is a downside to this for pollen – it sometimes gets eaten! Bees are the principal group of pollen consumers, and the most widely studied. For bees pollen grains are the principal source of non-liquid food; they collect and utilise pollen as brood food to rear their young. Pollen grains

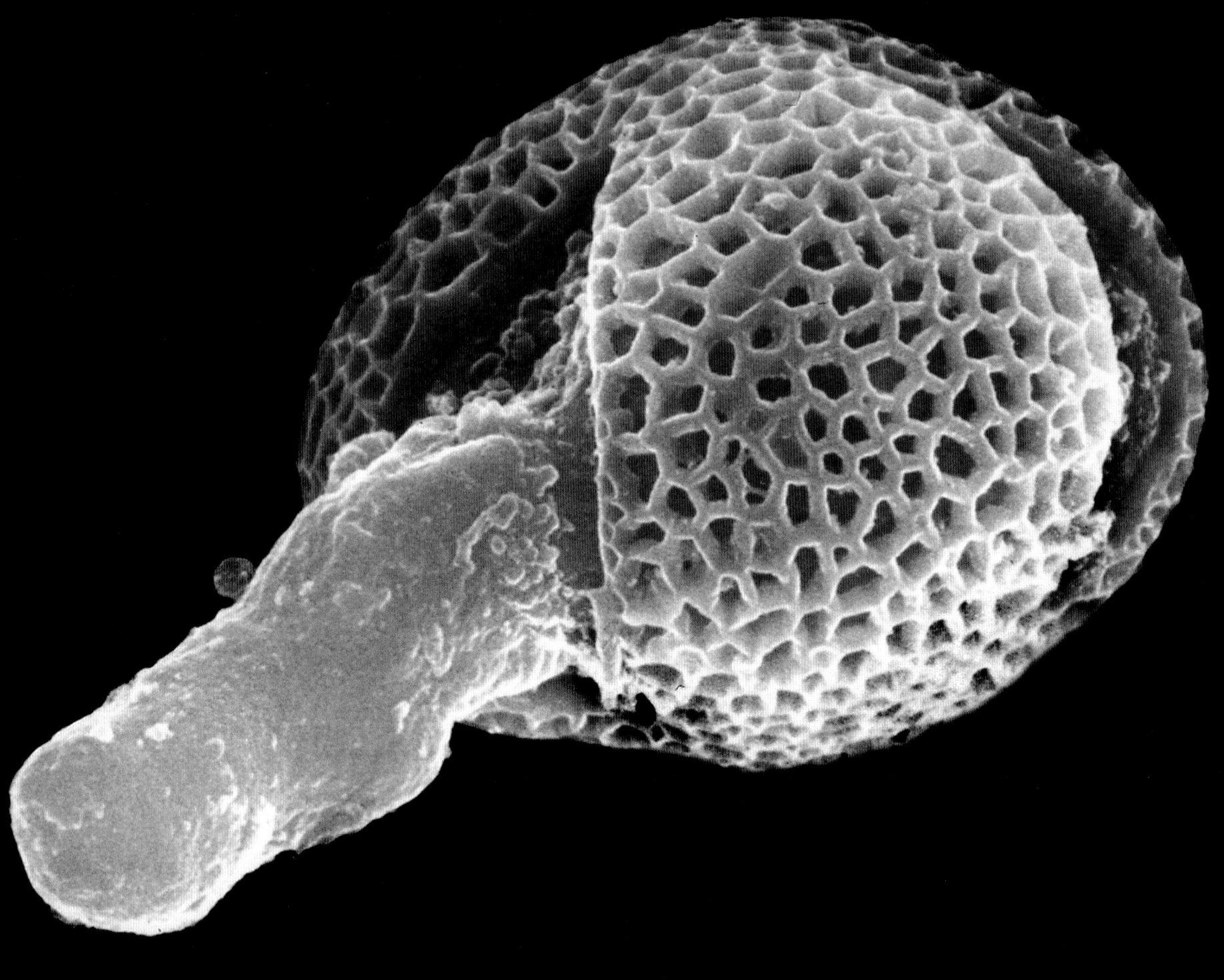

140

are also a regular source of food for many other species of insects, including beetles and flies. Some bats (suborder, Megachiroptera), as well as Humming Birds, and Australian Honey Eater birds eat pollen.

THE SEX ACT

At maturity the generative cell of the pollen grain undergoes a further mitosis, dividing to provide two sperm cells. This division happens either just before the pollen grains leave the anther or at any time after and up to just before the pollen tube reaches the embryo sac.

When a pollen grain lands on the surface of a stigma, recognition chemicals, contained on and within the surface of the pollen grain, will trigger a negative signal (sporophytic incompatibility) or a positive signal from the recognition chemicals contained within the surface of the stigma. A positive signal indicates that the pollen grain seems to be 'compatible', i.e. from the same species of plant.

Following successful acceptance by the stigma, usually in competition with hundreds, even thousands, of other pollen grains, instantly each pollen grain becomes involved in a fierce race to be first to deliver its sperm cells to an ovule – but first the sperm cells must get through one of the apertures in the pollen wall. This happens as follows: on contact the pollen grain absorbs moisture from the receptive surface of the stigma. The thick, inner, non-sporopolleninous layer of the pollen wall, the intine begins to expand below an aperture ('intine plug'), where it is even thicker. The aperture membrane splits open and the intine forces its way through this opening (analogous to icing being squeezed out of a syringe). The extruding intine – the pollen tube – surrounds the cellular material from the pollen grain containing the vegetative and sperm cells. The role of vegetative nucleus is not well-understood, but it is thought to be involved in the growth and development of the pollen tube. Pollen tube cells are the fastest growing of all plant cells,

Hamamelis mollis – Witch Hazel (Hamamelidaceae) – a germinating pollen tube emerging from an aperture. [CPD/SEM x 4000]

overleaf: *Prunus dulcis* – Almond (Rosaceae) – pollen grains germinating on sucrose-based agar medium. [SEM x 1000]. Image courtesy of Dr María Suárez-Cervera

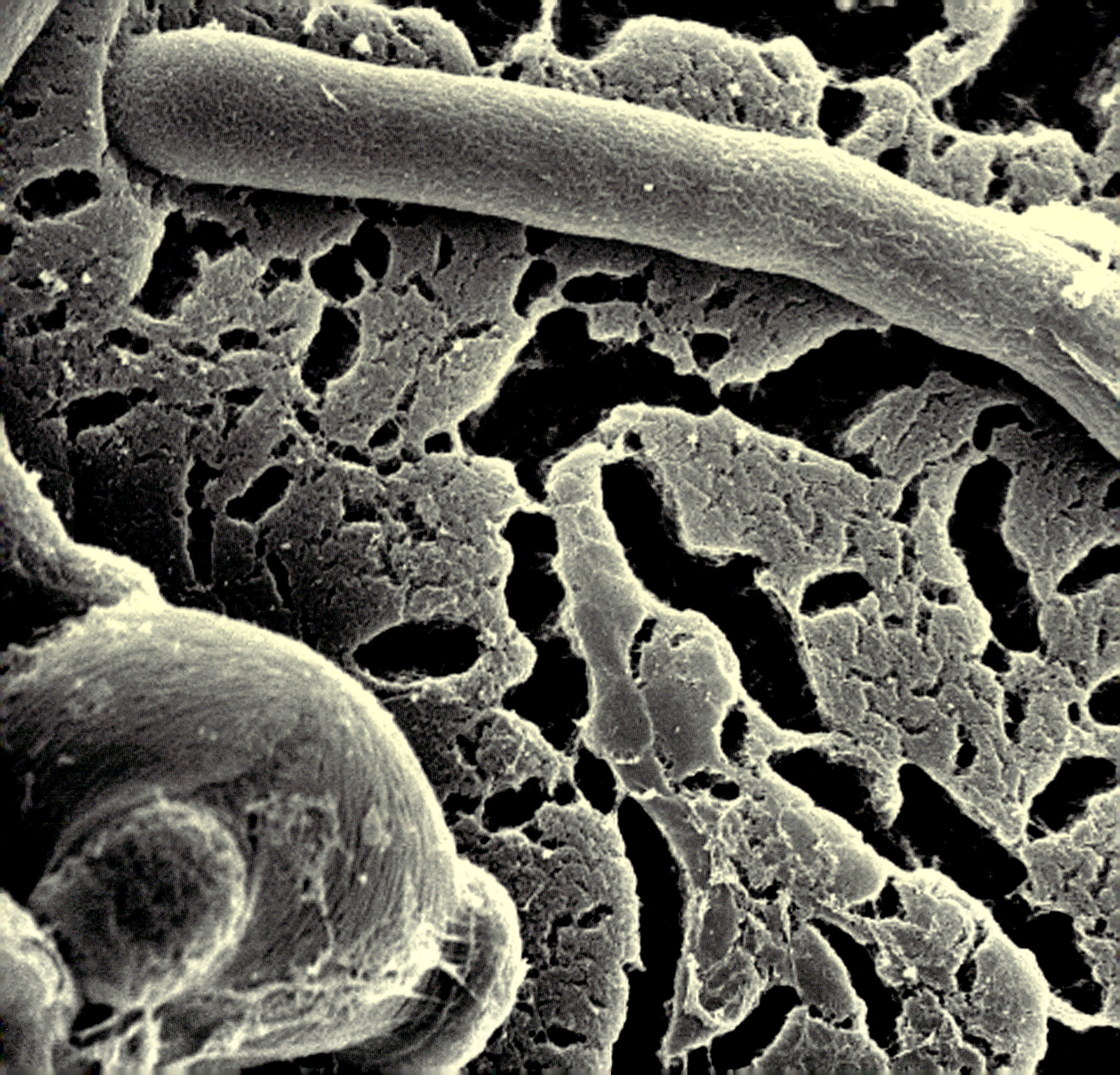

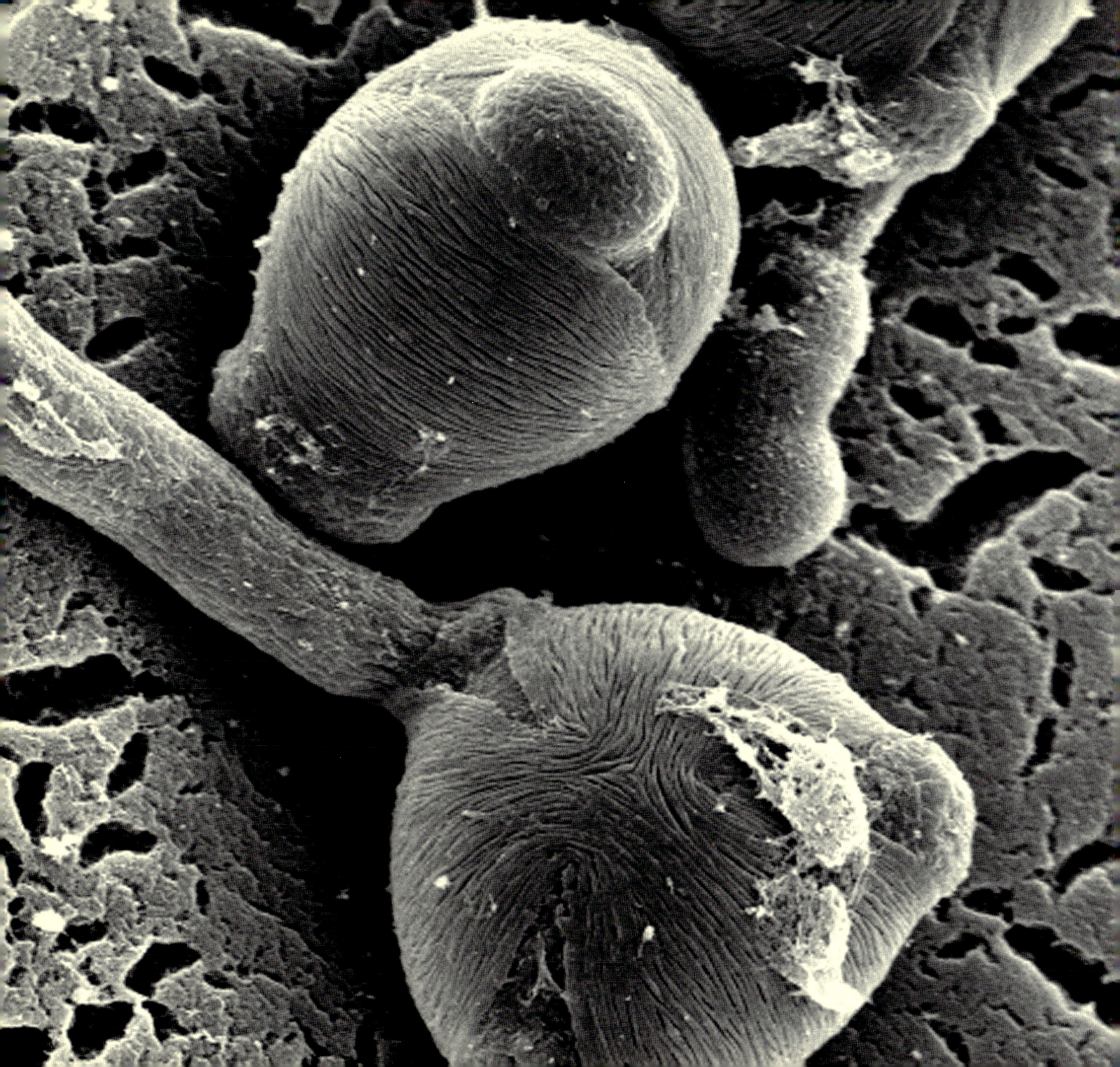

because they are competing against each other to be first to reach an ovule. The tubes develop rapidly, growing longer and thinner as they race down through the stylar canal to reach the ovary locule where the ovules are housed. The two sperm cells are carried along near the tip of the elongating pollen tube, with the vegetative nucleus just ahead. The germinating pollen tube also has to gain acceptance by the pistil as it travels through the stylar canal. If the pollen is not from the same species, it will be rejected before it reaches the ovule (gametophytic incompatibility), even though it has passed the sporophytic incompatibility test on the stigma surface. On arrival in the ovary the vegetative nucleus disintegrates, and the sperm cells are stripped of their cytoplasm, before continuing into the embryo sac of the ovule. Here one fuses with the ovum (egg nucleus) to form the embryo, and the other fuses with the polar nuclei to form the triploid primary endosperm nucleus, from which the seed storage tissue, the 'endosperm', will develop. This is the 'double fertilisation event', typical of flowering plants.

The beautiful outer casing of the pollen grain lies crumpled and discarded on the stigma surface, like festive wrapping paper after the gift has been removed, its function discharged. This is, of course, not the end of the story because now, following successful fertilisation by the pollen grain, the ovule can develop into a ripe seed, carrying with it the full chromosome complement necessary for growing into a new diploid plant.

Lunaria annua – Honesty (Cruciferae) – seed capsule

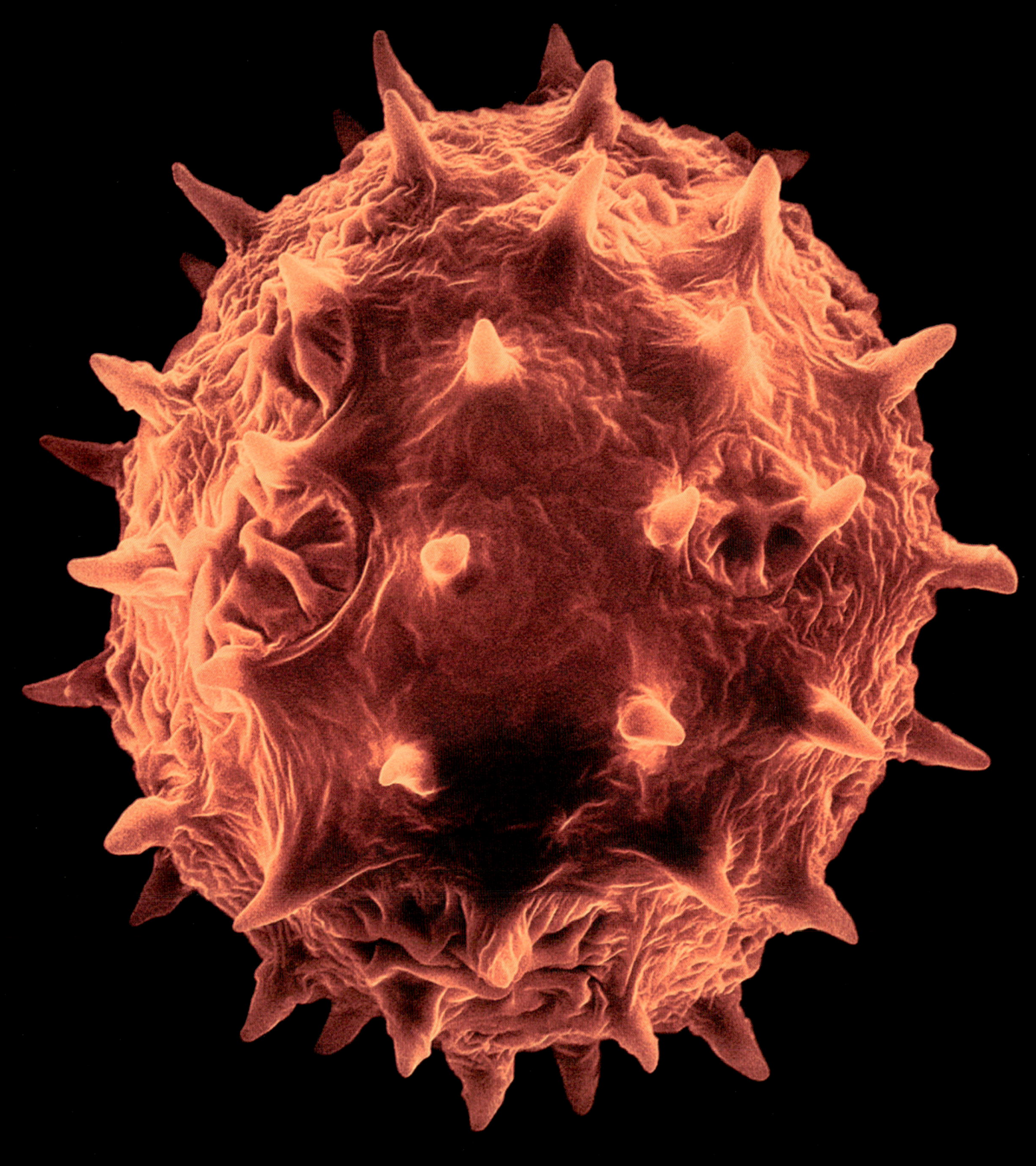

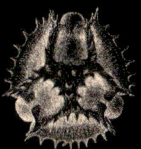

PICTURING THE INVISIBLE

Rob Kesseler

Anisodontea scabrosa (Malvaceae) --
pollen grain in natural condition [SEM x 1000]

Mallow
f. 14.
The spermatick Globulets in f. 13.

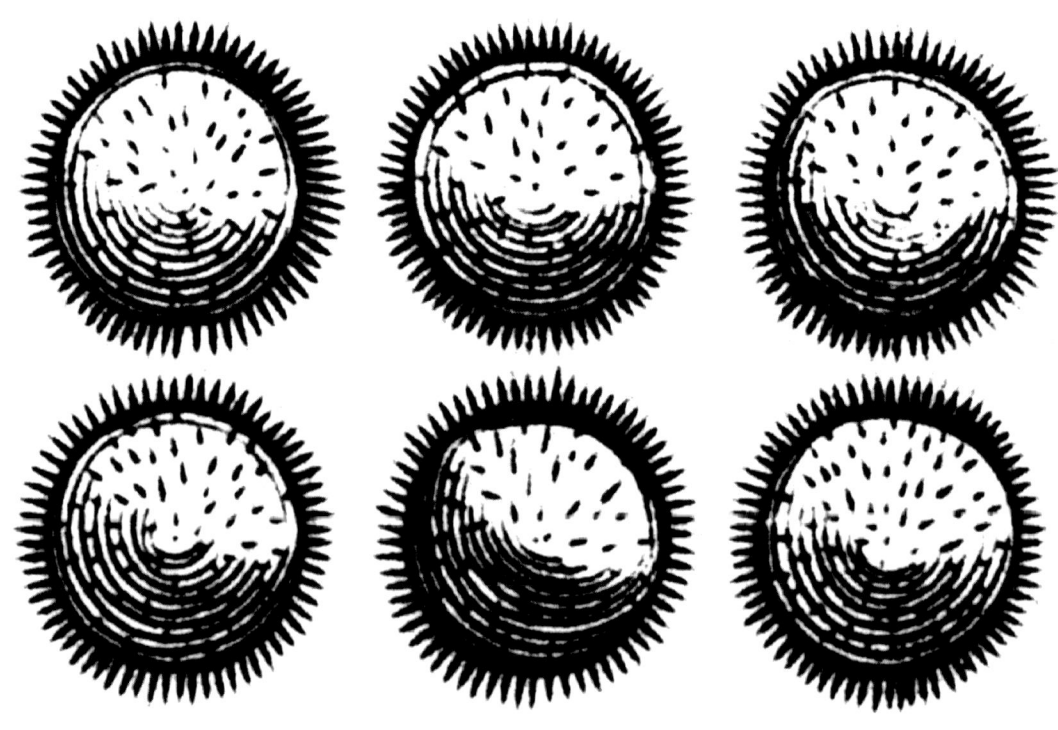

The art of pollen should perhaps be viewed within the broader context of flowers and their association with art. The urge to portray and understand the flowers and plants that surround us has a long and glorious history. They appear as decoration on the earliest pottery, and as a source of inspiration for the forms and structures of Greek architecture; they are celebrated in the flamboyant mastery of Dutch painting, and closely observed within the meticulous precision of botanical illustration. The creative outpouring of botanical decoration has served many needs and continues to migrate into every corner of our daily lives. Botanical images are eaten off, sat upon, slept under; they adorn our clothes and our walls. They have become powerful symbols that carry many messages, markers with which we retain contact with the natural world. it is hard to imagine a part of our lives they do not touch upon.

Historically, as the botanical sciences developed, and became increasingly sophisticated, so too did the means of representation, in a subtle fusion of observation and interpretation. From the seventeenth century the development of lenses and microscopes made possible the study of plant anatomy at levels beyond the scope of the human eye, providing botanists and artists alike with new subjects to work from. As early as 1676, in his rather racy botanical publication, *The Anatomy of Flowers, prosecuted with the bare eye, and with the microscope*, Nehemiah Grew (1641-1712), described in accurate detail the forms and functions of pollen from different plants. Appreciating their sexual significance he referred to pollen as "spermatic globulets", his line engravings set out in explicit detail the intimacies of plant genitalia. By 1735 sexual metaphor abounded, Carl Linnaeus (1707-78), described the corolla petals as "curtains of the nuptial bed". For Linnaeus's *Systema Naturae*, George Ehret illustrated the twenty-four sexual practices of plants, identified by the number of stamens and carpels. They were presented like miniature bouquets, laid out with the order implicit in the system of classification.

Nehemiah Grew (1682) 'spermatic globulets' – pollen grains of *Malva sylvestris* – Common Mallow (Malvaceae) – from *The Anatomy of Flowers Prosecuted with the bare eye, and with the Microscope*. [Courtesy of the Library, Royal Botanic Gardens, Kew]

page 150: Butterfly – *Polyommomatus icarus* – Common Blue (Lycaenidae) on *Onobrychis viciifolia* – Sanfoin (Leguminosae)

page 151: George Dionysius Ehret – detail of a Hibiscus with palmate leaves of 7 segments (*folus palmato-digitata Septemparitis* [sic]), (Malvaceae) flower, 1761. Watercolour. [Courtesy of the Library, Royal Botanic Gardens, Kew]

page 152: Walter Hood Fitch – *Rhododendron arboreum* var. *limbatum* (Ericaceae) 1862; note poricidal anthers. From the Arthur Church Collection of Botanical Drawings. [Courtesy of the Library, Royal Botanic Gardens, Kew]

page 153: *Rhododendron* cv. (Ericaceae) – stamens with poricidal anthers.

page 154: James Sowerby – a 'bizarre' Tulip. Plate IV from *Flora Luxurians*, 1789. [Courtesy of the Library, Royal Botanic Gardens, Kew]

page 155: Florist's Tulip showing the anthers and the tri-lobed stigma typical of Tulips.

151

154

In England, under the tutelage of George III, the early eighteenth century was a pivotal time for scholarship. It saw the emergence of a number of learned societies devoted to furthering ideas and exchanging knowledge within the arts and sciences. Originally housed within Somerset House, in 1874 these learned societies were relocated to Burlington House, off Piccadilly. The Society of Antiquaries, the Royal Society of Chemistry, The Geological Society, The Royal Astronomical Society, The Royal Academy and the

Linnean Society co-habited the buildings around the courtyard, an enlightened community, where intellectual cross-referencing became standard practice and a matrix for understanding the world.

There is a point of view, a contentious one, which is that botanical illustration has little to do with art. Aesthetic considerations are deemed inappropriate, anonymity of artistic expression an ideal, and beauty a pleasant but unimportant side effect. Apart from doing a disservice to the artists involved, one has only to look back over centuries of illustration to realise that this is a fallacious argument, an idealised impossibility. On the recommendation of Sir Joseph Banks, unofficial director of Kew Gardens from 1772 until his death in 1820, botanical illustrators like Ferdinand Bauer often accompanied plantsmen on global expeditions, equal partners in collecting, recording and disseminating information on plants from all over the world. Ferdinand's brother, Franz Andreas Bauer (1758–1840), like the best of botanical illustrators, possessed both artistic prowess and botanical insight; so much so that he received a handsome lifetime annuity from Banks to work from the living plant collections at Kew. Although his contemporaries paid scant regard to his scientific curiosity, his pioneering work on the systematic value of pollen morphology contributed greatly to the expansion of interest in this field. Before Franz Bauer's death there had already been a sudden burst of interest in pollen morphology; almost every botanist of note was intrigued by it. Furthermore, the rapid improvement in microscopes at the beginning of

below: Franz Bauer, *Passiflora caerulea* pollen, from his collection of illustrations, *Epidermis floris. Pollen grains, Monstrosities.* [Courtesy of the Natural History Museum, London]

opposite: *Passiflora caerulea* – Passion Flower (Passifloraceae) – pollen in natural condition [SEM x 1000]

overleaf left: Carl Julius von Fritzsche, hand-coloured engraving of *Alcea rosea* (Malvaceae), from *Ueber den Pollen*, 1837. [Courtesy of the Library, Royal Botanic Gardens, Kew]

overleaf right: *Malva sylvestris* – Common Mallow (Malvaceae) [SEM x 4800 – acetolysed]

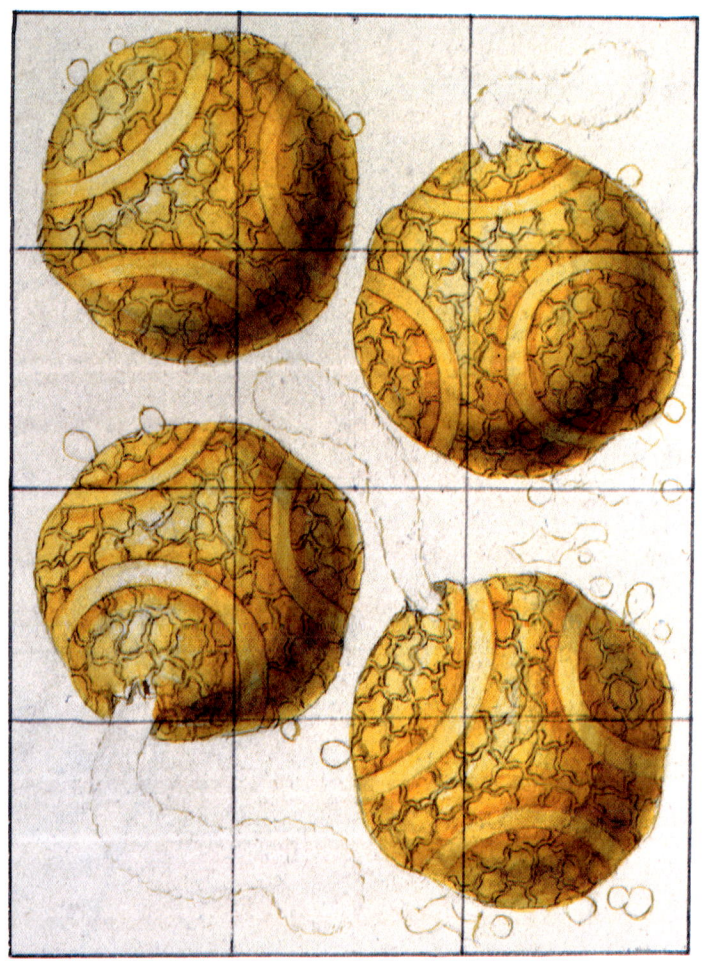

Pollen: The Hidden Sexuality of Flowers

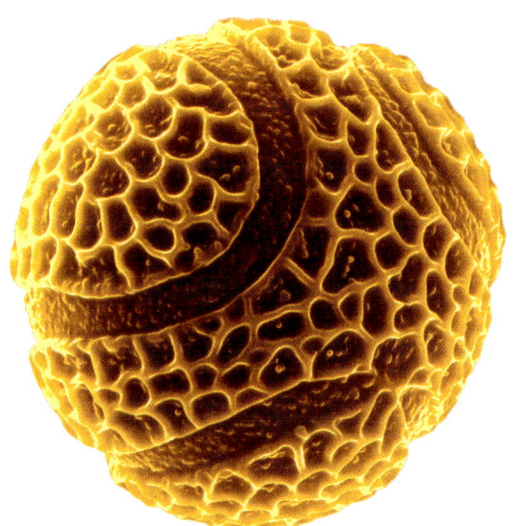

the nineteenth century enabled Franz Bauer to produce highly detailed studies of many pollen types, resulting in a collection of detailed drawings and watercolours of pollen grains, now held in the Botanical Library of the Natural History Museum, London.

The emerging technology of microscopes encouraged new research into the subject, contributing to a rapid expansion of information and images. In 1837 chemist and amateur botanist Carl Julius von Fritzsche (1808-1871) published *Ueber den Pollen*, a detailed study of pollen morphology in which he revealed the diverse, complex and highly ornamental structures of pollen grains. His studies displayed a high degree of detailed observation of the distinctive and diverse forms of pollen grains; but, whether out of the desire to make taxonomic distinctions, or for reasons of imposing order through stylistic convention, it appears he could not resist the temptation to improve on their symmetries, tidying up their arrangement of spines and apertures.

This was a seminal moment in the history of botanical science, for at the very moment when Carl Julius von Fritzsche was publishing his hand-illustrated work, he was also beginning to experiment with the newly invented process of photography. Using materials supplied from London he was able to make simple contact prints, photogenic drawings of leaf silhouettes, showing structural details of veins. Forty years of experimentation by a collection of gentlemen scientists across northern Europe laid the foundations for this new art form. Thomas Wedgwood (1771–1805), the son of potter Josiah Wedgwood (1730-1795), was experimenting with ways to create mechanical images using light and silver compounds, and had he not died young, may well have gone on to play an even greater role in the development of photography. In France Joseph Nicéphore Niépce (1765-1833), scientist and lithographer, experimented with ways to capture images directly from nature onto lithographic plates and, in 1827, he produced his first *heliographs*. Convinced of their benefits to botanists, not only as a record but also as a means of providing

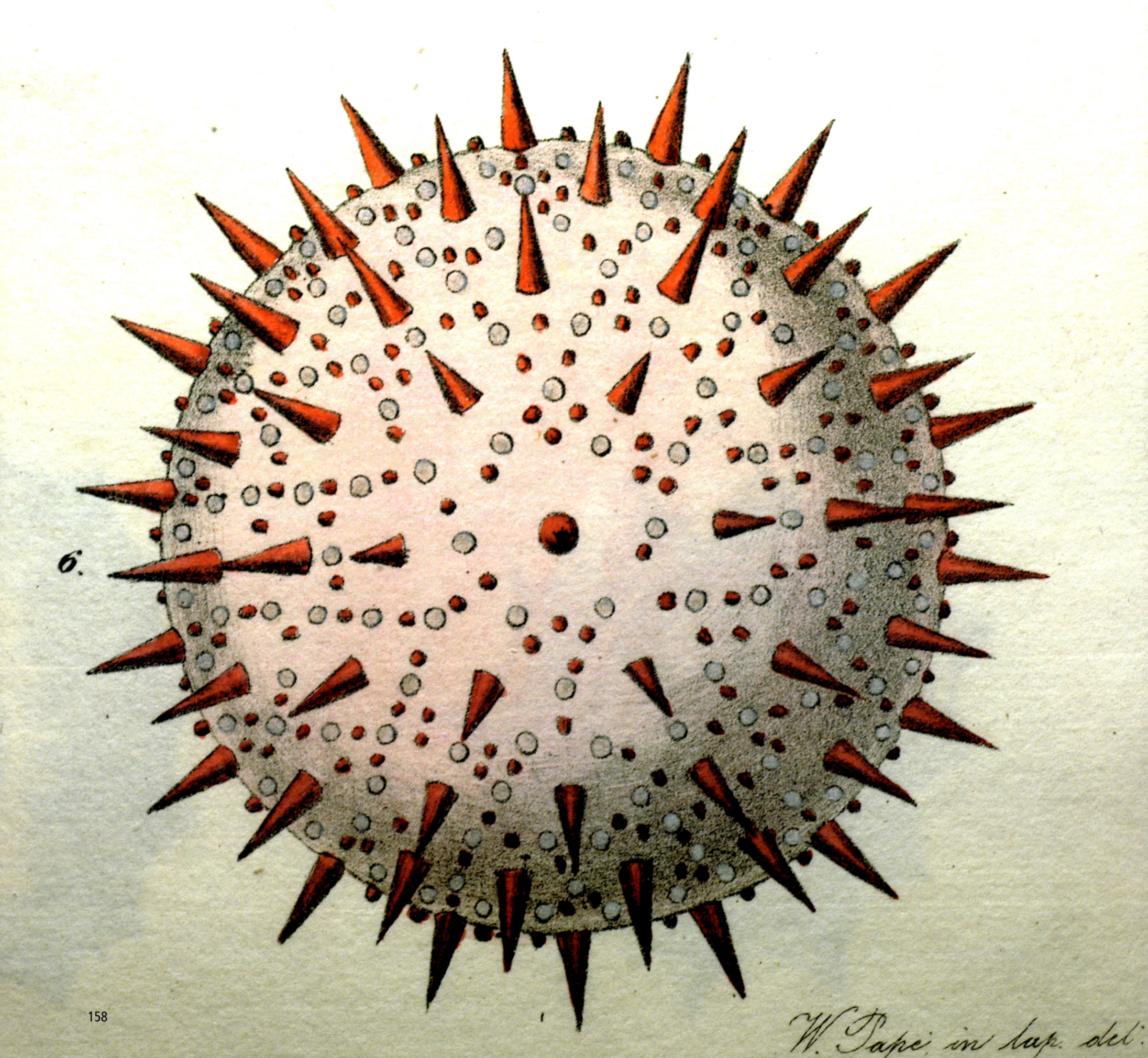

6.

W. Pape in lap. del.

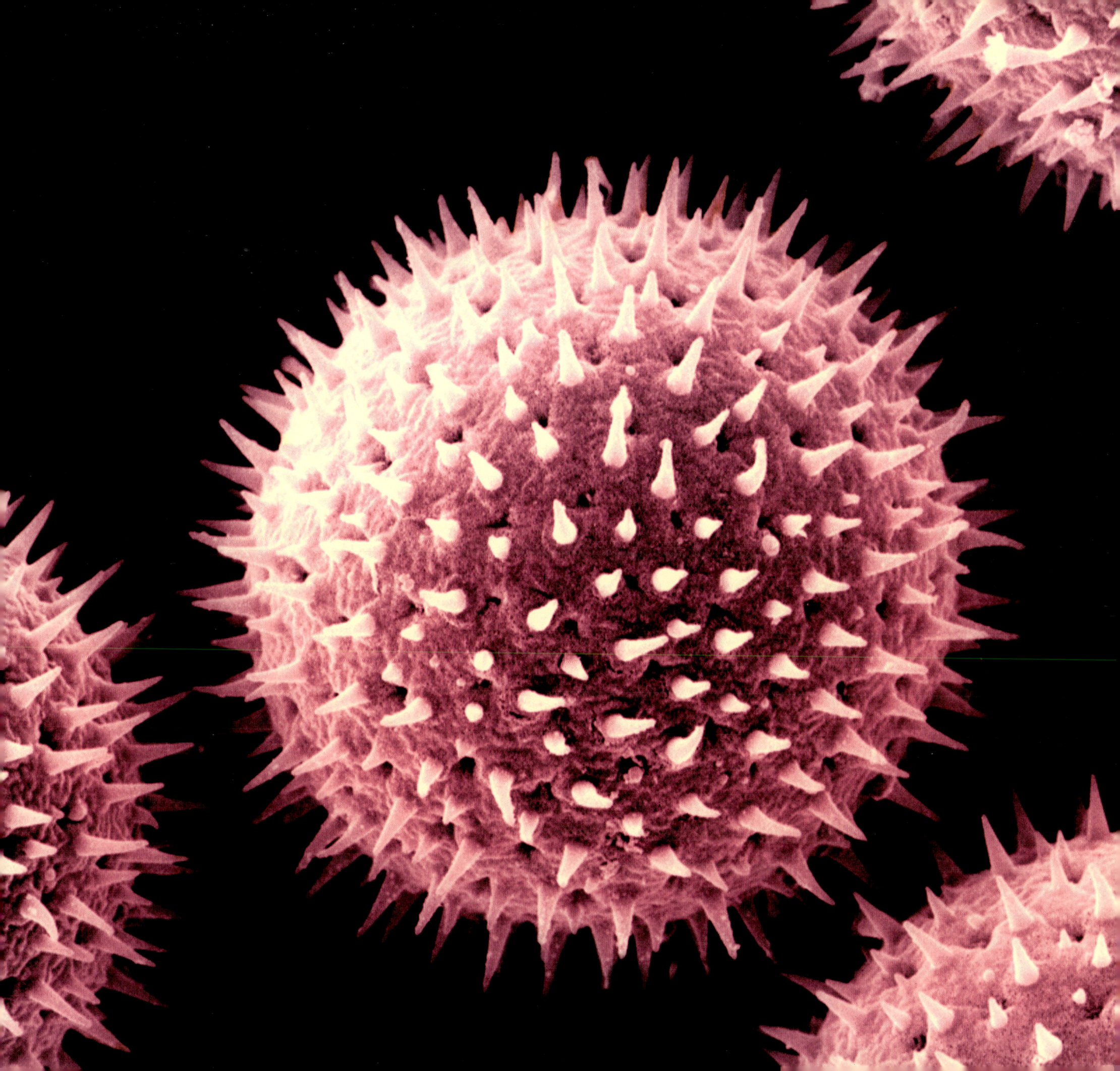

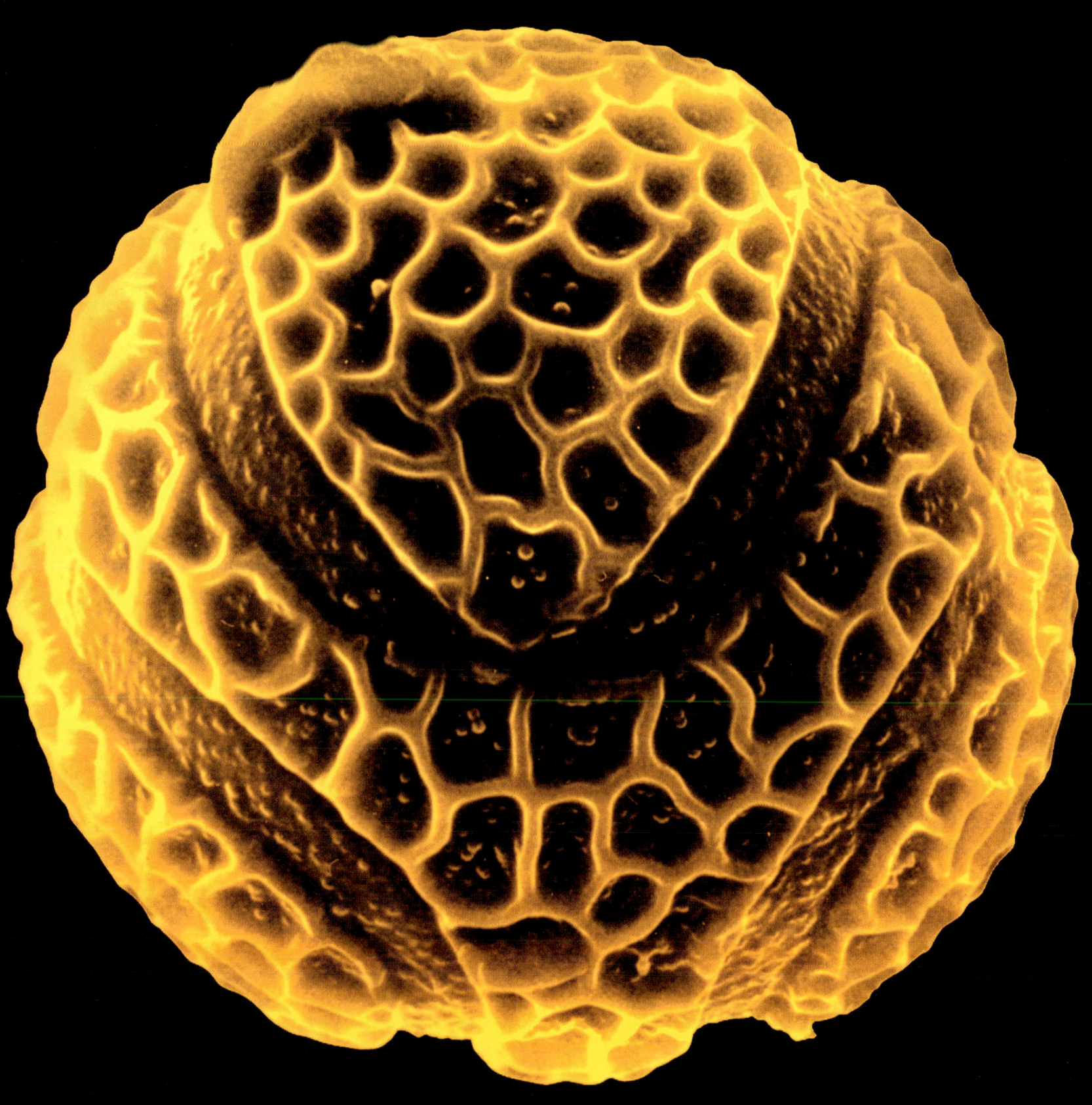

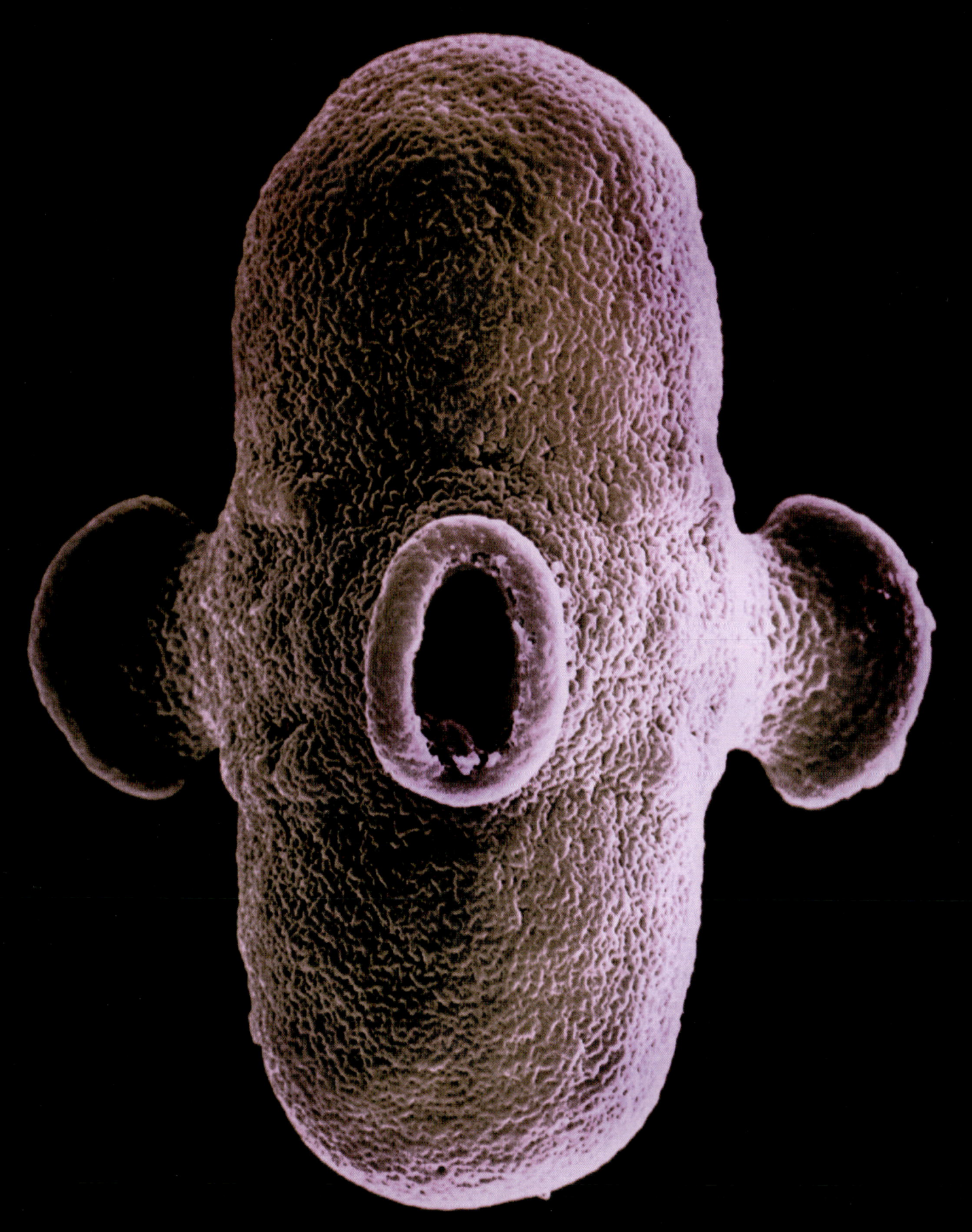

George Dionysius Ehret – detail of Granadilla (*Passiflora quadrangularis*). Watercolour illustration,1757. [Courtesy of the Library, Royal Botanic Gardens, Kew]

opposite: *Passiflora caerulea* – Passion flower

duplicates for wider distribution, he travelled to England in an attempt to present his discoveries to the scientific community. With the exception of Franz Bauer at Kew, who proved to be a strong champion of his work, most of the people who could have been influential were either indisposed or out of town. Despite Bauer's enthusiasm for his discoveries, Joseph Nicéphore Niépce found little other support and, dismayed, returned to France. Here he renewed a relationship, made prior to his departure for England, with Louis Jacques Mandé Daguerre (1787–1851), known at that time as the inventor of the diorama. Within three years Niépce was dead but the invaluable information gained through their partnership enabled Daguerre to continue his developmental work and, in 1839, he produced the first reliable photographic images, 'daguerreotypes'.

In Britain, there was already a well-established practice for using optical devices to facilitate the recording of very small, or microscopic, details of the plants and flowers being studied. Botanists such as Sir John Herschel (1792-1871), Anna Atkins (1799-1871) and William Henry Fox Talbot (1800-1877) were all versed in the art of the camera obscura and camera lucida. However, Fox Talbot realised that although these devices did enable drawing they could not always compensate for his lack of ability as a draughtsman. Reflecting on the possibility of being able to fix natural images to the page, he started to experiment by coating sheets of paper with solutions of salt and silver nitrate. Placed under glass, with a leaf sandwiched between, the paper was exposed to sunlight for fifteen minutes. Wherever the light struck the silver was transformed into silver chloride, creating a shadowgram of the leaf.

In the space of a few years the pioneering work of Daguerre and Fox Talbot was consolidated. The expertise and materials became easily available throughout the world, resulting in the proliferation of highly sophisticated photographic images of every conceivable subject. Within the realm of botanical science the birth of photography heralded a new dawn, an

opposite: Anna Atkins – *Sargassum bacciferum* – a brown free-floating seaweed from the Sargasso sea; from *Photographs of British Algae*,1843. Cyanotype photogramme. [Courtesy of the Library, Royal Botanic Gardens, Kew]

Sir John Herschel – *Experiment 512*: four leaves – water fixed,1839. Photogenic drawing negative. [Courtesy of the National Museum of Photography, Film & Television/Science & Society Picture Library]

Pollen: The Hidden Sexuality of Flowers

Sargassum bacciferum

inseparable union of art and science driven by their respective needs and aspirations, which complemented the tradition of botanical illustration and extended its ability to communicate the subject to wider audiences. Pollen was among the subjects to receive attention from this newly established science, for example in the studies of John Samuel Slater (1850-1911), a Civil Engineer in Calcutta. After his retirement in 1904, Slater returned to England where, among his numerous scientific interests, he took up the study of pollen. The pollen grains were photographed onto glass plate negatives for magic lantern projection. After his death his collections of pollen images were bequeathed to the Royal Botanic Gardens, Kew.

Study of the natural world at microscopic level was not, of course, limited to plants. Towards the end of the nineteenth century Ernst Haeckel (1834-1919) Professor of Zoology at Jena, and polemical advocate of Darwinism, described over four thousand different species of marine radiolaria. With their astonishing diversity of form and structural ornamentation (at times uncannily similar to those of pollen) it is not surprising that Haeckel suggested that protoplasm may possess an "inherent artistic drive". The zoomorphic patterns and arabesques of the marine life forms reproduced in his *Kunst-Formen der Natur* (1899-1904) echo the fluid and ornate forms found in the architecture and decorative arts of the Jugendstil and Art Nouveau movements. His observations found material form in the work of the sculptor Anthony Cragg (b. 1949) almost a century later, in a collection of sculptures, '*forminifera*'. Cragg describes the information we gain about the natural world as, "important because in this huge storeroom lie the keys to essential processes and explanations of our existence. The application of natural principles and models results in a flora of utilitarian objects, environments and events that are subjugated to functionalism."[1]

At the beginning of the twentieth century, at the Botanic Gardens in Oxford, botanist and artist Arthur Harry Church (1865-1937) worked on a detailed examination of the internal sexual organs of flowers which for Church, "…were not decorative additions to Western culture; they were machines ensuring successful reproduction in plants."[2] The resulting paintings portrayed a clinical textbook observation of flower anatomy, but so suggestive in form, explicit in detail, sensual in colour and voyeuristic in presentation that by default these floral images verged on the erotic; anticipating the later flower paintings of Georgia O'Keeffe (1887-1986) and the photographs of Robert Mapplethorpe (1946-1989). There is a nice irony in knowing that Church executed his paintings at a time when the legacy of Victorian prurience still held sway and, perhaps, this contributed to his own view that they were merely by-products of his research with little artistic value.

By-products take different forms and, in Berlin during the same period that A.H. Church was active in Oxford, Karl Blossfeldt (1865-1932) was working with similar fascination and attention to detail to develop one of the most influential pieces of photographic documentation of plant forms of the twentieth century. Blossfeldt was an amateur photographer, artisan and lecturer, who was inspired to create a record of the diversity of plant forms as a teaching aid for future designers in the Berlin School of Arts and Crafts, where he taught. Shunning exotic flowers like orchids and lilies, "Blossfeldt frequently gathered his plants on country tracks, on railway embankments, and in other similarly 'proletarian places'. For him it was often the plants generally and unjustly denigrated as weeds that had the most fascinating forms".[3] Up until this time there had been little macroscopic work in photography but, by developing his own cameras, he was able to photograph flower heads, buds and seeds in dramatic close-up, revealing the minutest surface details in hyper-reality. Laid out against a plain background, the symmetries of form were presented with a clarity that transcended their diminutive scale, rendering them deceptively monumental. The collected images, when published in 1928, made him an overnight success, provoking curiosity and fascination,

John Samuel Slater – a pollen grain of *Mirabilis jalapa* – Marvel of Peru (Nyctaginaceae), 1907 (LM).
Plate negative for magic lantern slide.
[Courtesy of the Palynology Unit, Royal Botanic Gardens, Kew]

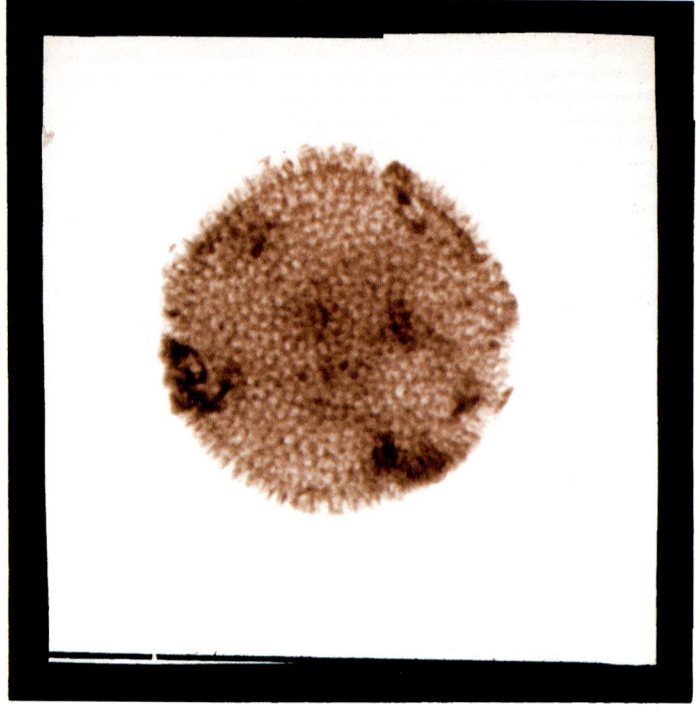

John Samuel Slater – pollen grain of *Geranium nodosum* –
Knotted Cranes-bill (Geraniaceae), 1907 (LM).
Plate negative for magic lantern slide.
[Courtesy of the Palynology Unit, Royal Botanic Gardens, Kew]

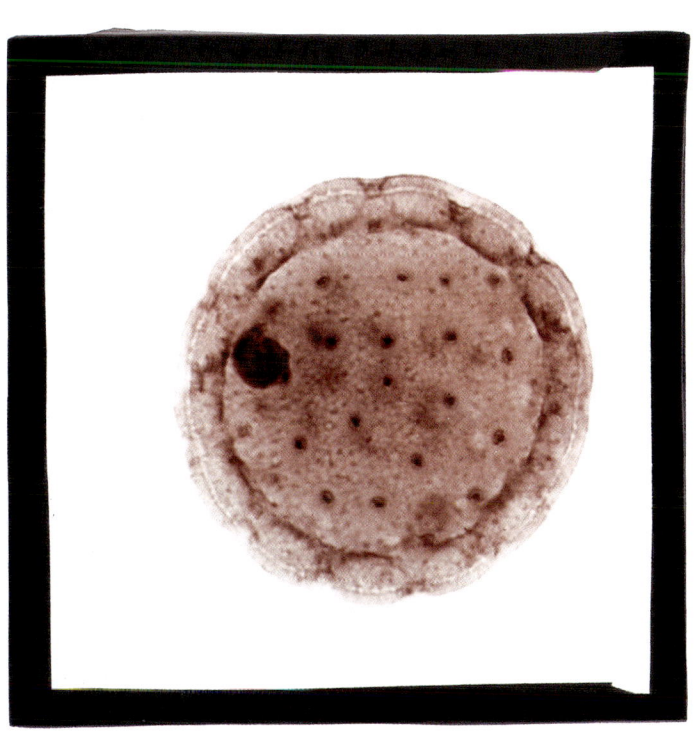

demonstrating the widespread appeal of his subject within the burgeoning fascination for nature studies sweeping Germany at that time. In many ways echoing the rigour of plant systematics, it is noteworthy that such a simple approach to content and construction was able to influence such a diverse spectrum of artistic activity, and laid the foundations for a new attitude towards objectivity. A line of photographic descent can be traced from the almost surrealist works of Man Ray (1890-1976) and Edward Weston (1886-1958), through to the serial photographs of industrial buildings by Bernd and Hiller Becher. As with Haeckel, architects were also drawn to Blossfeldtt's work but this time not for its decorative form but rather for the structural complexities that were revealed.

The eighteenth and nineteenth centuries also saw developments in other printing processes, in particular the advance from copperplate etching to the more sophisticated colour lithography, which enabled the publication of collections of botanical prints, which found an eager and receptive audience. During the same period, the growing ceramics and textile industries were quick to capitalise on the popularity of botanical art; this resulted in a proliferation of elaborate dinner services, as well as luxurious fabrics decorated with copies from botanical collections. In this way botanical illustrators had not only made an important contribution to plant identification, they were also instrumental in creating the affectionate place plants hold within society. Exotic or familiar, images of plants quickly became celebrated as signifiers of well-being, confirming the role of botanical illustration, not just within the scientific community but also across society at large.

From the time of the Great Exhibition at the Crystal Palace in 1851, there emerged clearly defined sets of laws relating to the use of plants as a source for design, a grammar of ornament. Within the decorative and applied arts, Owen Jones (1809-1874), John Ruskin (1819-1900), and later William Morris (1834-1896), all laid out their theories for the ways in which pattern and ornament

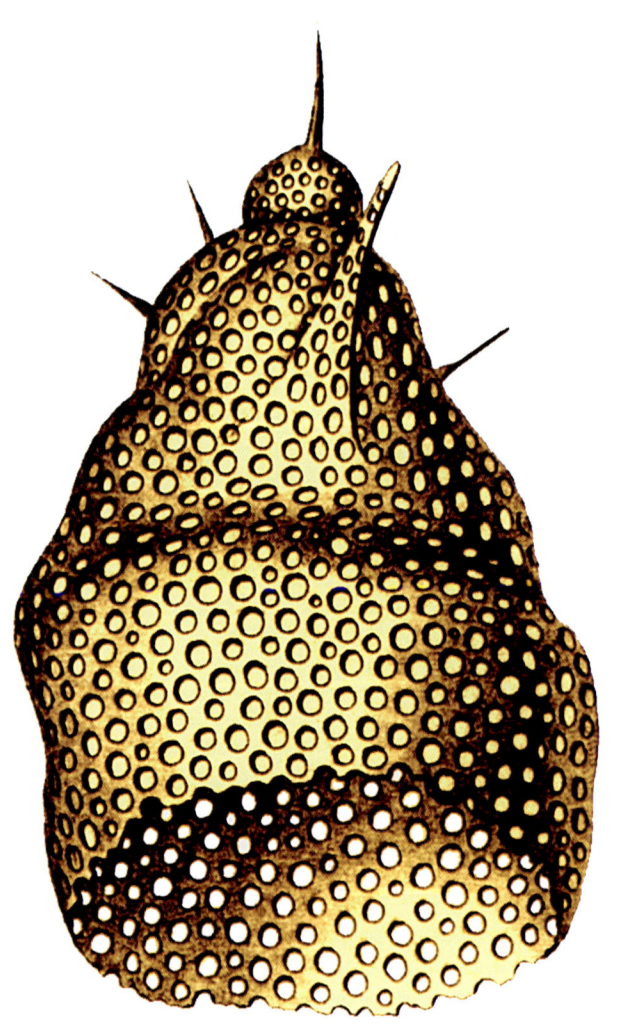

Ernst Haeckel – detail of a plate from *Die Radiolarien*, Berlin, 1862

opposite: Anthony Cragg – *Envelope* (1998), bronze
[Courtesy Anthony Cragg and the Lisson Gallery, London]

relating to plants should be abstracted: stylisation over representation. Christopher Dresser (1834-1904), arguably the first modern product designer, while a student at the Government School of Design at Somerset House, worked directly from plants sent down weekly from Kew Gardens. These early studies became the inspiration for his designs, and spanned the whole field of decorative arts: furniture, carpets, ceramics, silverware, ironwork, wallpaper, stained-glass. The depth of his knowledge was recognised in his appointment as 'Professor of Botany Applied to the Fine Arts', for the Department of Science and Art in South Kensington. In 1860 he was awarded an honorary doctorate from the University of Jena for his contribution to botanical science. The use of plants as a direct source of inspirational design was to continue as the Arts and Crafts movement evolved to rise to the challenge of what they saw as a declining public taste for industrial artifacts towards the end of the nineteenth century.

In time, the prevailing climate for these forms of stylisation gave way to the more naturalistic approach of Art Nouveau and Jugendstil, a more fluid and reflective interpretation derived from the forms of nature. As the apparent sensuousness of this approach was edged out by the hegemony of Modernism, and with the notion of ornament as crime, proposed by the Austrian architect Adolf Loos, (1870-1933), there appeared little room for the exotic excesses of flowers. It seems ironic that, in the light of Modernism, the new 'machine age' should fail to take advantage of emerging technological developments, which could have pushed back the boundaries of creative exploration even further within the botanical sciences. As a result spectacular images resulting from plant microscopy remained very much confined to the botanical domain, with artists finding only occasional, tantalising glimpses within the pages of scientific journals, and failing to build on the promising advances of early photography. One of the few exceptions to this was the French photographer, Laure Albin-Guillot (1879-1962) who, supported by her scientist husband, explored the aesthetic potential of microphotography to develop a new

170

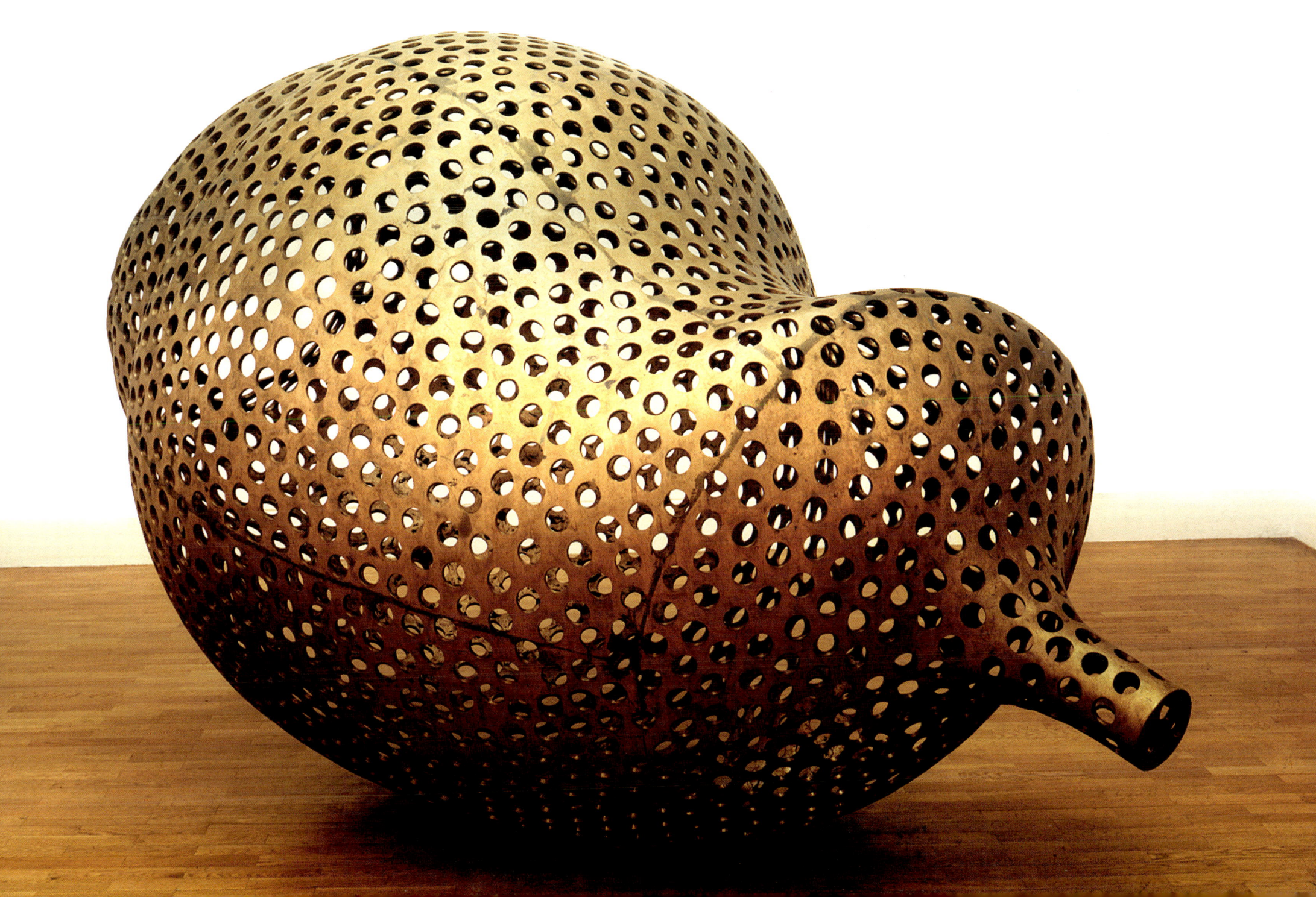

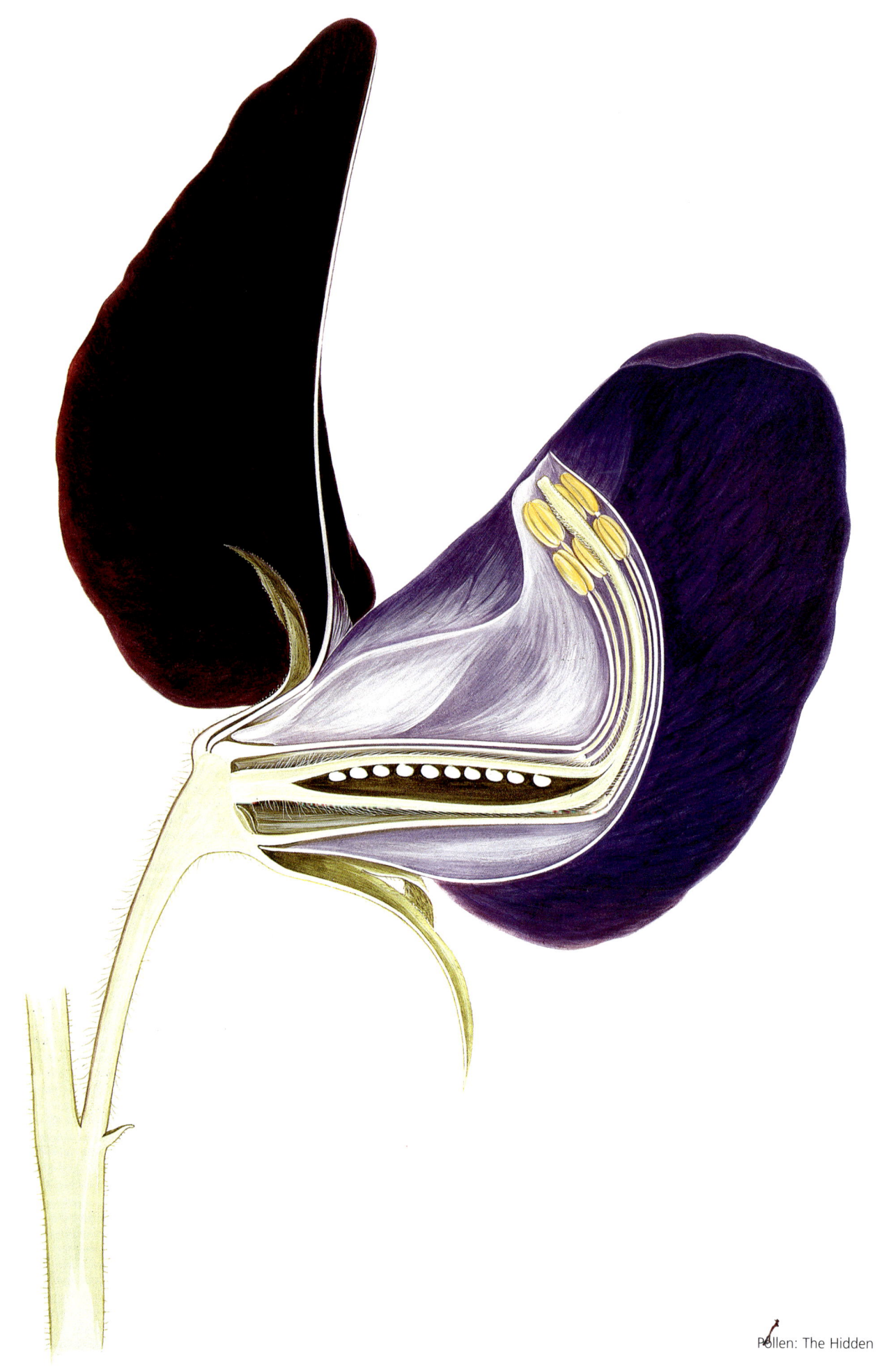

vocabulary of natural design, *Micrographie décorative*. Using metallic inks to print her photogravures, Albin-Guillot transformed microscopic plant and mineral structures into exotic geometric patterns, richly decorative, and with an abstract ornamentalism reminiscent of the prevailing Art Deco style.

Botanical illustrators continue to play an important part in plant identification. Their work, selective and detailed representations, and masterpieces of composition, can highlight characteristics of plants, drawing attention to particular features in ways that are more difficult for the camera. However, by the twentieth century the balance, at the microscopic level, was shifting. Camera lenses and microscope optics were being improved rapidly, yet the photographic images were still much less informative than drawings of direct observations of pollen, and other subjects, seen under the microscope. This is exemplified by the fine pen and ink drawings of highly magnified pollen by the Canadian biologist and chemist Roger Philip Wodehouse (1889–1978) culminating in his definitive textbook *Pollen Grains* published in 1935. The tradition of drawing pollen grains was to be carried through into the 1960s by the Swedish botanist, Gunnar Erdtman (1897-1973). In his diagrammatic work on pollen morphology he illustrated not only whole pollen grains but also surface details of the exine, as well as sections through pollen walls. His textbook *Pollen Morphology and Plant Taxonomy*, published in 1952, rapidly became an indispensable reference for students and professionals, and is still in use today. Before Erdtman's death, remarkable advances were being made in the development of electron optics. During the 1940s and 1950s these advances began to have an impact on microscopy, and microscopic imaging in the biological sciences. This post World War II era embraced a feeling that hand-drawn illustration was no longer appropriate, the hand no substitute for the sophisticated detail possible using modern cameras attached to light or electron microscopes. The advent of scanning and transmission electron microscopy put an even greater distance between viewer and object

Arthur Harry Church – *Lathyrus odoratus* – Sweet Pea (Leguminosae), 1903. Watercolour. [Courtesy of the Natural History Museum, London]

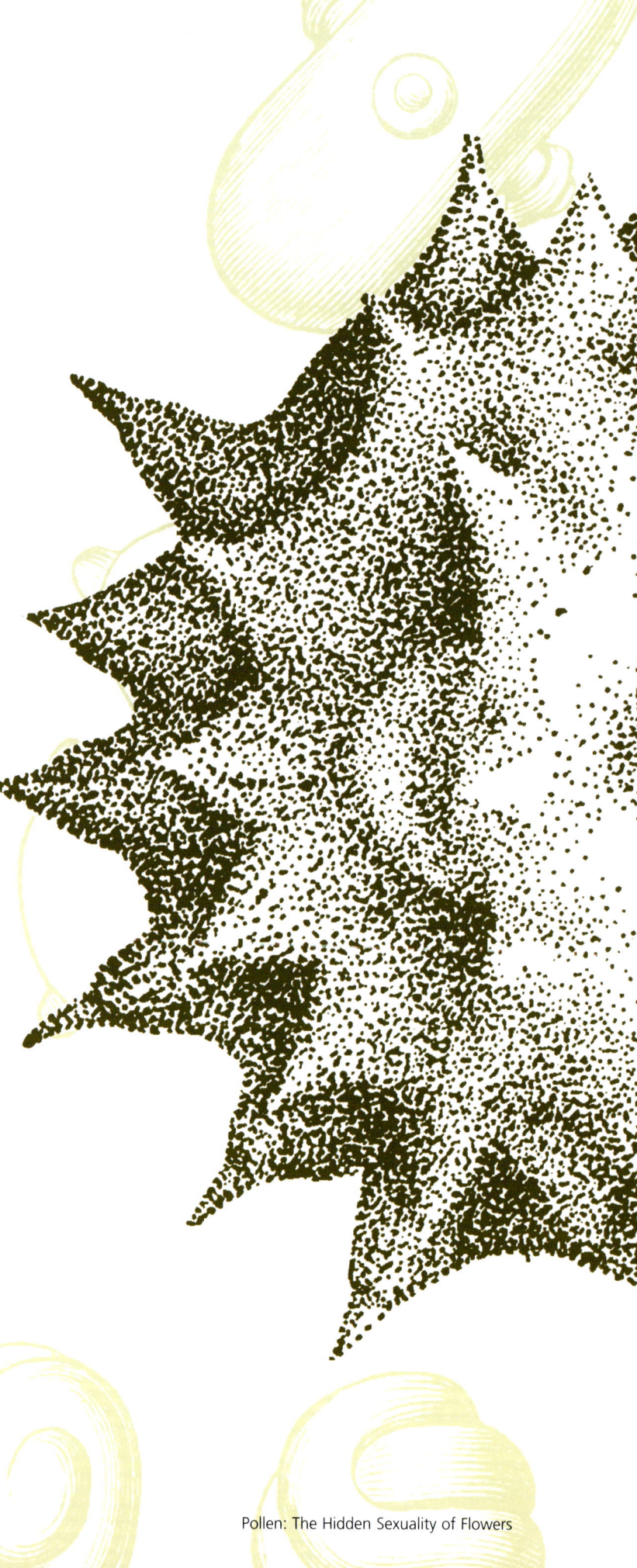

than the light microscope, a distance mediated by the operator through the use of photography, its very nature leading to an anonymity of authorship and cementing the idea that the camera cannot lie.

When viewing the rich history of botanical representation, it is clear how the appeal of such powerful imagery filters into the wider artistic community, to be dispersed beyond the borders of its original botanically-oriented intention. Over time the huge body of work by botanical artists has become a major primer for wider artistic inspiration and abstraction, and its influence continues to reach out to a highly receptive audience. However, as research techniques became more sophisticated during the early part of the twentieth century, there appeared to be a hiatus in simultaneous collaboration between artists and scientists, a separation that is now being restored to the benefit of both. It could be argued that the point at which a camera was attached to a microscope was the point at which the visualisation of botanical science went underground, out of reach to all but a privileged minority. Progressively, specialism can lead to isolationism and, for the greater part of the twentieth century, both the artistic and scientific communities have at times been guilty of working behind the closed doors of laboratories and studios, secure within their own practice, looking down the wrong end of the telescope (or in this case microscope), protective of their own positions and detached from society. Recently there has been a reversal in this trend with the realisation that more broad-minded support can provide the impetus and encouragement needed for artists and scientists to develop their ideas in a mutually creative atmosphere. Museums, galleries and arts agencies have been expected to play a more proactive role in widening participation through the development of projects that place artists in new situations. A growing number of Sci-Art initiatives have emerged to foster new debates around science, and to encourage collaboration and communication. It is not that art and science are on convergent paths, rather that they operate in parallel universes, but the

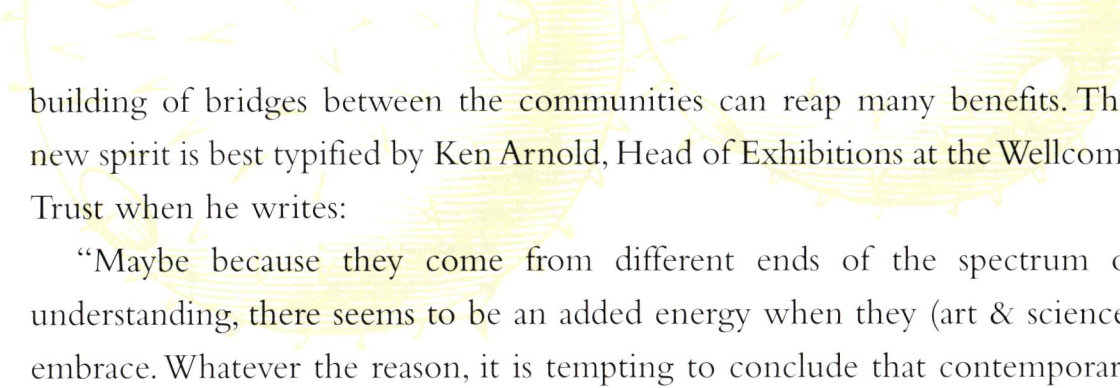

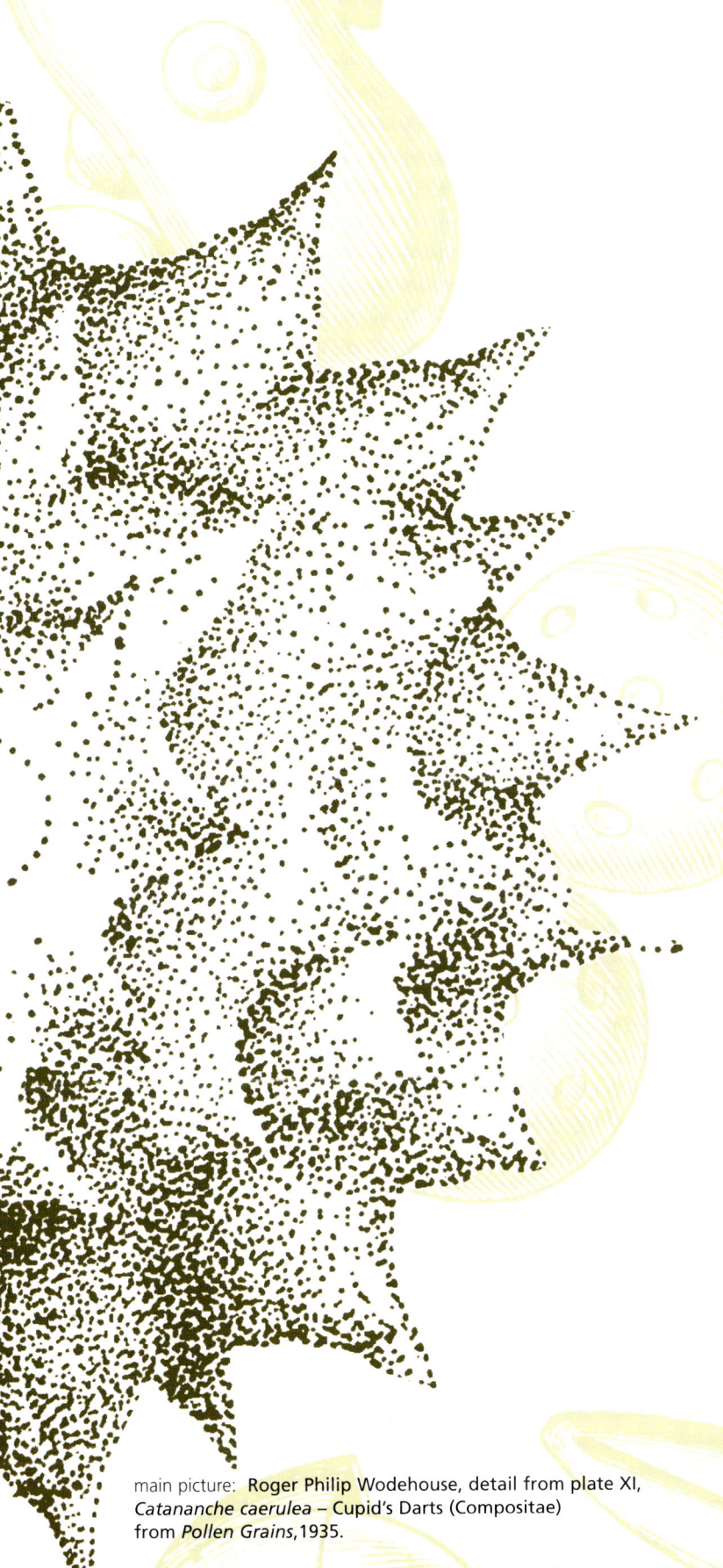

building of bridges between the communities can reap many benefits. This new spirit is best typified by Ken Arnold, Head of Exhibitions at the Wellcome Trust when he writes:

"Maybe because they come from different ends of the spectrum of understanding, there seems to be an added energy when they (art & science) embrace. Whatever the reason, it is tempting to conclude that contemporary science and art have found gaps in each other that required filling."[4]

Increasingly, artists have been pushing at an opening door. Receptive to new opportunities, they have become adept at expanding their field of enquiry and sphere of operation, increasingly aware of the possibilities of creating a wider audience for their work. Artists understand that in this new climate of cooperation they have potential access to areas normally the domain of a privileged few; in this way they have become an alternative conduit through which the of life sciences may be examined. Working from the premise that knowledge shared is knowledge multiplied, it is now not uncommon to see artists working alongside neuroscientists, biotechnologists, automotive designers, sound engineers, zoologists and animal behaviourists.

It is tempting to think that the benefits are loaded on the side of the artist; one might well ask, what is in it for the scientist? There are often naïve and misplaced expectations of the potential outcomes of art and science collaborations, which take a narrow view of the diversity of practice within the respective fields. Opportunities for scientific discoveries are hard won, so what role is there here for the contemporary artist? Scientists, caught in a system which is intensely pressured and based on results may, of necessity, take a narrow view of their research and have little time to develop its value beyond the needs of their peers and clients. The more free-wheeling artist may not understand the science in depth, and the scientist may not immediately appreciate the philosophy of the artist. Nevertheless, meaningful collaboration between the two disciplines can reveal areas of scientific practice and discovery,

main picture: **Roger Philip** Wodehouse, detail from plate XI, *Catananche caerulea* – Cupid's Darts (Compositae) from *Pollen Grains*,1935.

background: A.Kerner & F.W.Oliver, illustration of pollen grains from *The Natural History of Plants*, 1903.

extending them beyond what would have been their normal audience. There has also been a convergence of technologies; the digital revolution has had a dramatic effect on both artists and scientists, who now share many of the same tools. In the studio and the laboratory visualisation happens as much on screen as on paper. Hand-drawn and photographic images merge, are manipulated, transformed, transmitted and, at the highest resolution, can be output in minutes. Electronic technology has become the medium through which a shared platform across disciplines has been able to develop. This technology has proliferated so fast across all disciplines that it has rapidly become the common language of the specialist and non-specialist.

Nevertheless, it is a powerful, and enabling language which has considerable impact on our ability to communicate complex information and ideas. In an information-rich culture, people are encouraged to expect access to information bases on topics that affect their daily lives. Therefore, it is increasingly important for scientists to develop a professional approach to the visual interpretation of their work to non-scientific audiences as well as in their own community.

The power of electron microscopy, the versatility of image manipulation packages, and the remarkable qualities of reproduction from high resolution printers, digital projectors and plasma screens mean that the resultant images have a clarity and strength which calls into question the need for any artistic intervention at all. One might speculate that the role of the botanical artist is in danger of being confined to the visible world. However, this would be to ignore the role of the artist in interpreting and translating the new imagery; to reflect the many ways in which the imperceptible parts of nature, such as pollen, have a profound but unnoticed impact on our lives.

Notes

1 Anthony Cragg, "Vantage point", *Art Monthly*, 1988
2 David Mabberley, *Arthur Harry Church, The Anatomy of Flowers*, Merrell, 2000
3 Hans Christian Adam, *Karl Blossfeldt*, Prestel, 1999
4 Ken Arnold, "Science and Art: Symbiosis or just good friends?", *Wellcome Trust News Supplement,* 2002

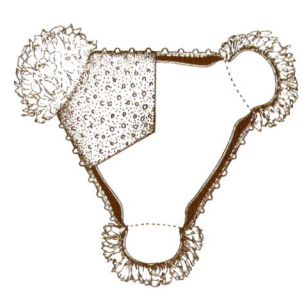 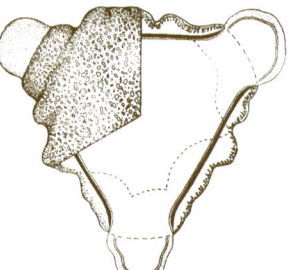

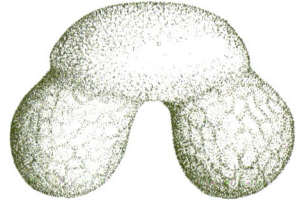 

top: Gunnar Erdtman, pollen grains from figures 214: top left: *Xylomelum angustifolium;* top right: 212 – *Hakea ruscifolia* (Proteaceae) from, *Pollen Morphology and Plant Taxonomy,* 1952 [Courtesy of Almqvist & Wiksell, Stockholm]

above: Roger Philip Wodehouse – detail from plate II, above left: *Pinus nigra* – Corsican Pine (Pinaceae); above right: *Pherosphaera fitzgeraldii* (Podocarpaceae), from *Pollen Grains: Their Structure, Identification and Significance in Medicine* (1935)

opposite: Roger Philip Wodehouse – detail from plate XII, *Barnadesia berberoides* (Compositae) from *Pollen Grains: Their Structure, Identification and Significance in Medicine,* 1935

PIXILLATED POLLEN

ROB KESSELER

*Nature must be 'put to the torture' and made to yield
its reluctant secrets to the astute investigator*

Francis Bacon
(1561 – 1626)

Nenga gajah – Pinang Palm (Arecaceae) –
pollen grain [SEM x 2000 – acetolysed]

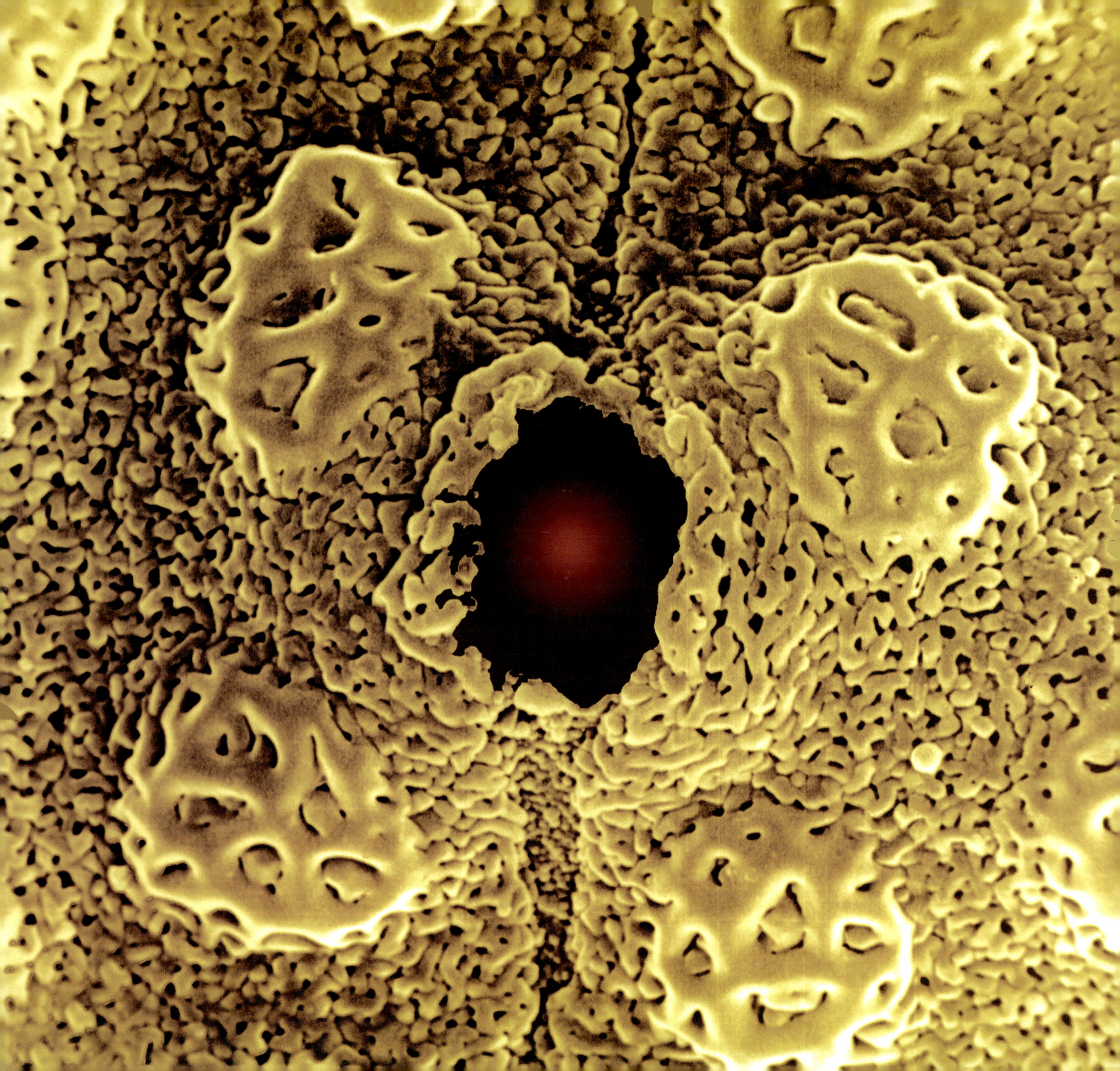

Sweet is the lore which nature brings;

Our meddling intellect

Mis-shapes the beauteous forms of things;

-- We murder to dissect.

Enough of science and of art;

Close up these barren leaves;

Come forth, and bring with you a heart

That watches and receives.

William Wordsworth
Extract from *The Tables Turned,*
1798

Justicia anisotoides (Acanthaceae) –
pollen grain [SEM x 1300 – acetolysed]

Nearly two centuries separate the contrasting views of Bacon and Wordsworth, and the tensions that existed between the Rationalist and the Romantic. Two hundred years on little has changed; the sentiments encompassed remain as topical, and opinion just as divided. Cognisant of the fears and fascinations of a society that views nature with a contradictory mixture of voyeuristic sentiment and environmental responsibility, artists and scientists continue to *dissect* nature, driven by professional orthodoxies and individual preoccupations. Divergent views prevail and the nature of plants is no exception. In one way or another scientists and artists working with plants, dissect, examine and analyse, modify and even transform their subjects. Their findings are presented through diverse channels of communication, contributing not only to an ever growing body of knowledge that will preserve and sustain diversity but also, by extension, to the debates surrounding our natural resources and their potential as a commodity for commercial exploitation and cultural consumption.

From the fifth century Byzantine Codex, *Aniciae Julianae* (the oldest illuminated version of the writings of Dioscorides regarding the medicinal properties of plants and minerals) through seventeenth century Dutch flower paintings, to the sexually-charged flower photographs of Robert Mapplethorpe in the late twentieth century, artists have been understandably aware of the visual potency of the material they work with. Images of plants have a magnetic attraction: informing, provoking thought, delighting and

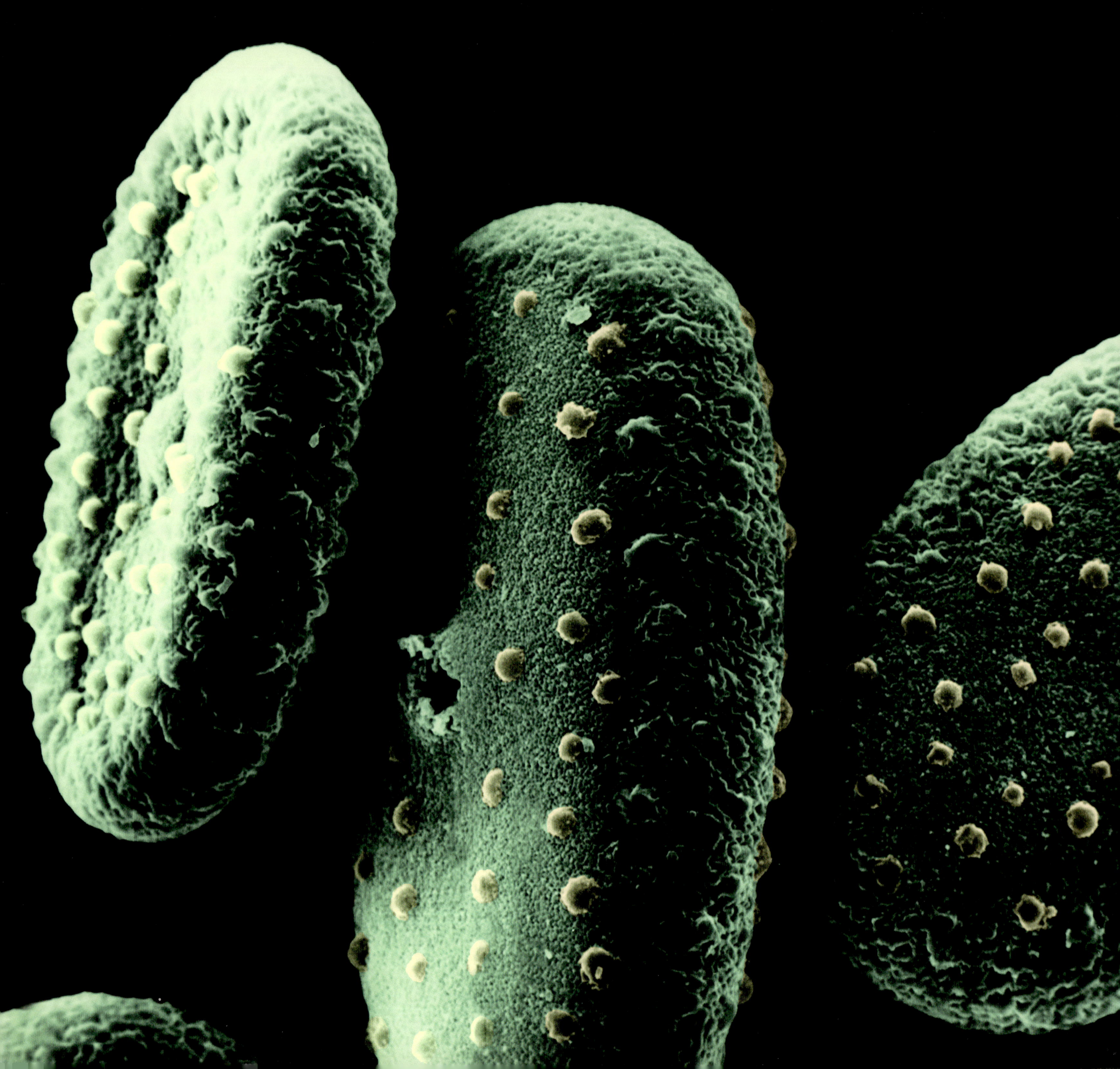

Macrorungia pubinervia (Acanthaceae) –
pollen grain [SEM x 1200 – acetolysed]

seducing us. Botanical imagery seems to succeed unfailingly in capturing new audiences. Here, details of plants, realised with the power of scanning electron microscopy and the digital camera extend a tradition that relies on the close collaboration of botanists and artists working as mutual partners towards the common goal of botanical description. The digital world we occupy provides a fertile breeding ground for art-science initiatives.

Following the development of the light microscope by Robert Hooke (1635-1703) and Anton van Leeuwenhoek (1632-1723), its potential for discovering the natural world was soon realised and, motivated by the visual potency of the material examined, it did not take long for the microscope to be put to uses other than for strict scientific enquiry. In 1726 as a form of popular entertainment, Martin Frobenius Ledermüller (1719-1769) put on shows in Germany known as *Augenergötzungen* (eye pleasure). Within a darkened room, glass containers filled with water were placed before his solar microscope, projecting spectacular images of the micro-organisms usually invisible to the naked eye.

Today 'eye pleasure' might be translated as 'eye candy', a derogatory term indicative of a society saturated with images designed for quick, easy gratification. However, it is a clumsy term that devalues the act and experience of looking; a phrase that fails to recognise the power of certain images to provoke the retinal response which precedes any cultural mediation. Notions of beauty and the sublime hover uneasily in a post-modern society saturated by persuasive images that have been stripped of their original aura. Our awareness has been blurred and our experience diluted by endless layers of cultural appropriation, as we experience the natural world through the eyes of others.

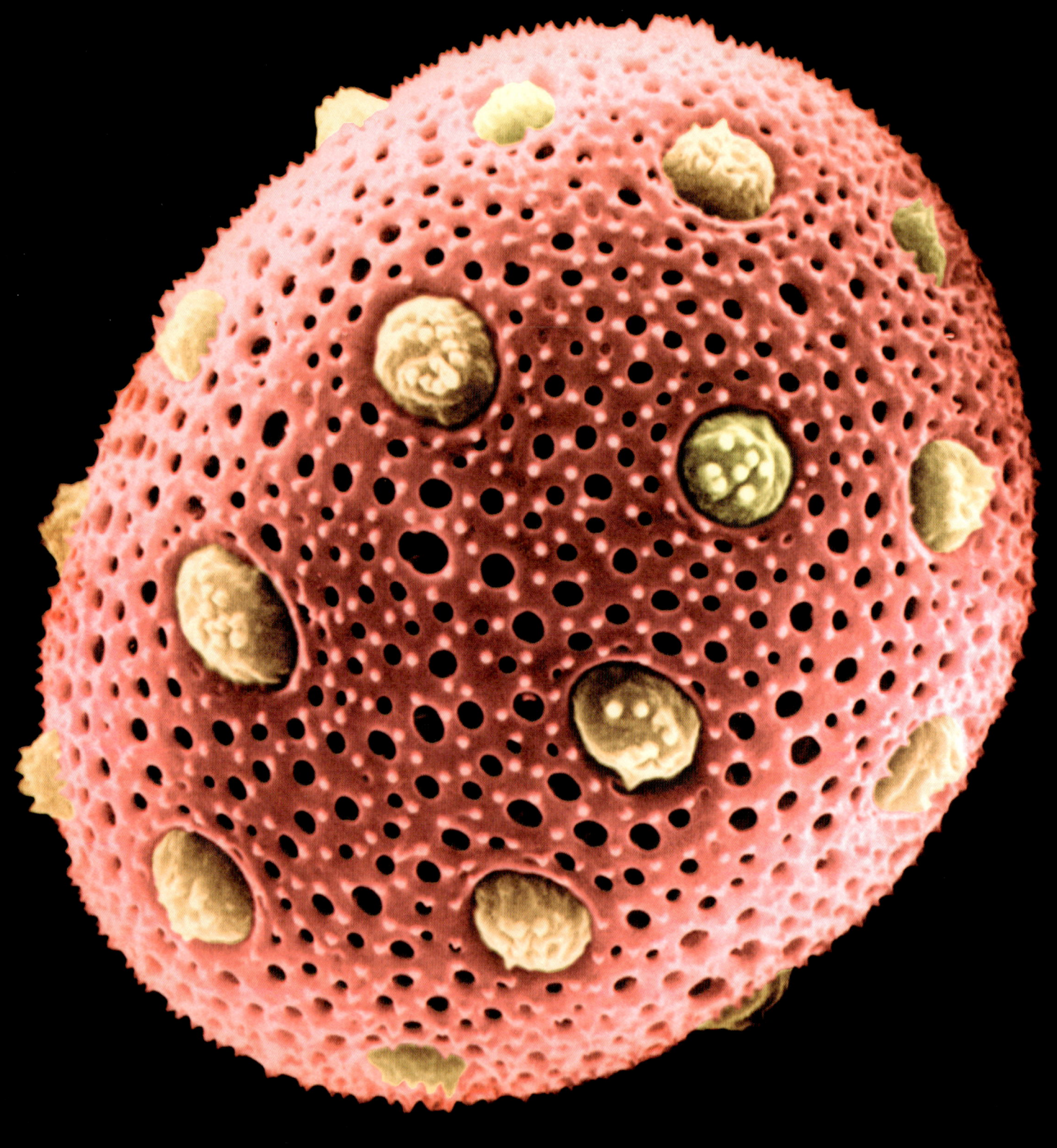

Perhaps we have lost the habit and the pleasure of looking for its own sake, as a first step towards a deeper understanding.

A detailed study of flower anatomy reveals unimaginable diversity. Moving below the surface, the minute and astonishing variety of form, structure and surface topology of pollen opens up yet another layer of visual and scientific complexity, a world invisible to most people. We may endlessly dissect but beauty, like infinity, is mercurial. That something as small as pollen is also so vital for the continued diversity of plant life is remarkable; that it is also beautiful adds immeasurably to its fascination.

Early collectors and artists often travelled great distances and endured much hardship in seeking out, recording and bringing back spectacular specimens of previously unknown plants from around the world. Today perhaps the journey is more sedentary: flowers become the continents; microscopes and computers the navigation aids for a new voyage of discovery. Throughout the history of plant cultivation, horticulturalists have drawn upon scientific knowledge to create new flower varieties; they are surrogate midwives to spectacular blooms and new colour breaks. Artists likewise want a hand in this act of botanical procreation, transforming an exotic fusion of scientific knowledge and artistic interpretation into a personal *phytopia*.

Silene dioica – Red Campion (Caryophyllaceae) – pollen grain [SEM x 2000]

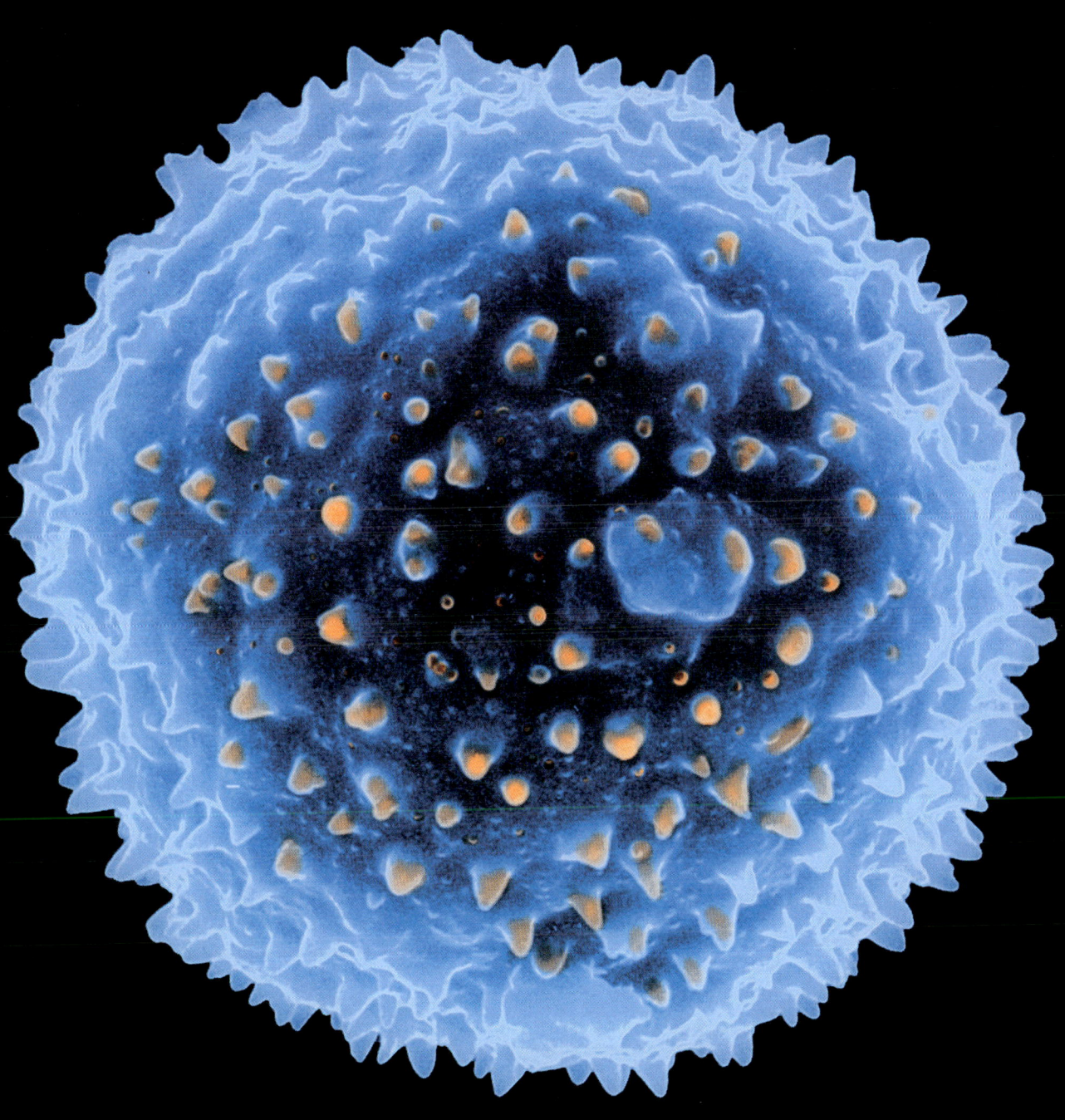

Meconopsis grandis – Himalayan Blue Poppy (Papaveraceae) –
pollen grain, semi expanded condition. [SEM x 1500]

opposite: *Meconopsis grandis* – Himalayan Blue Poppy (Papaveraceae)
– open flower; note the numerous stamens typical of poppy flowers.

Eustoma grandiflorum – Lisianthus (Gentianaceae) –
interior of flower, showing three of the five stamens, with open
anthers releasing pollen. In the centre the stigma at the top of the
style is dusted with pollen. At the base of the style is a 5-locular ovary.

opposite: *Eustoma grandiflorum* – Lisianthus (Gentianaceae)
pollen grains – dehydrated condition [SEM x 600]

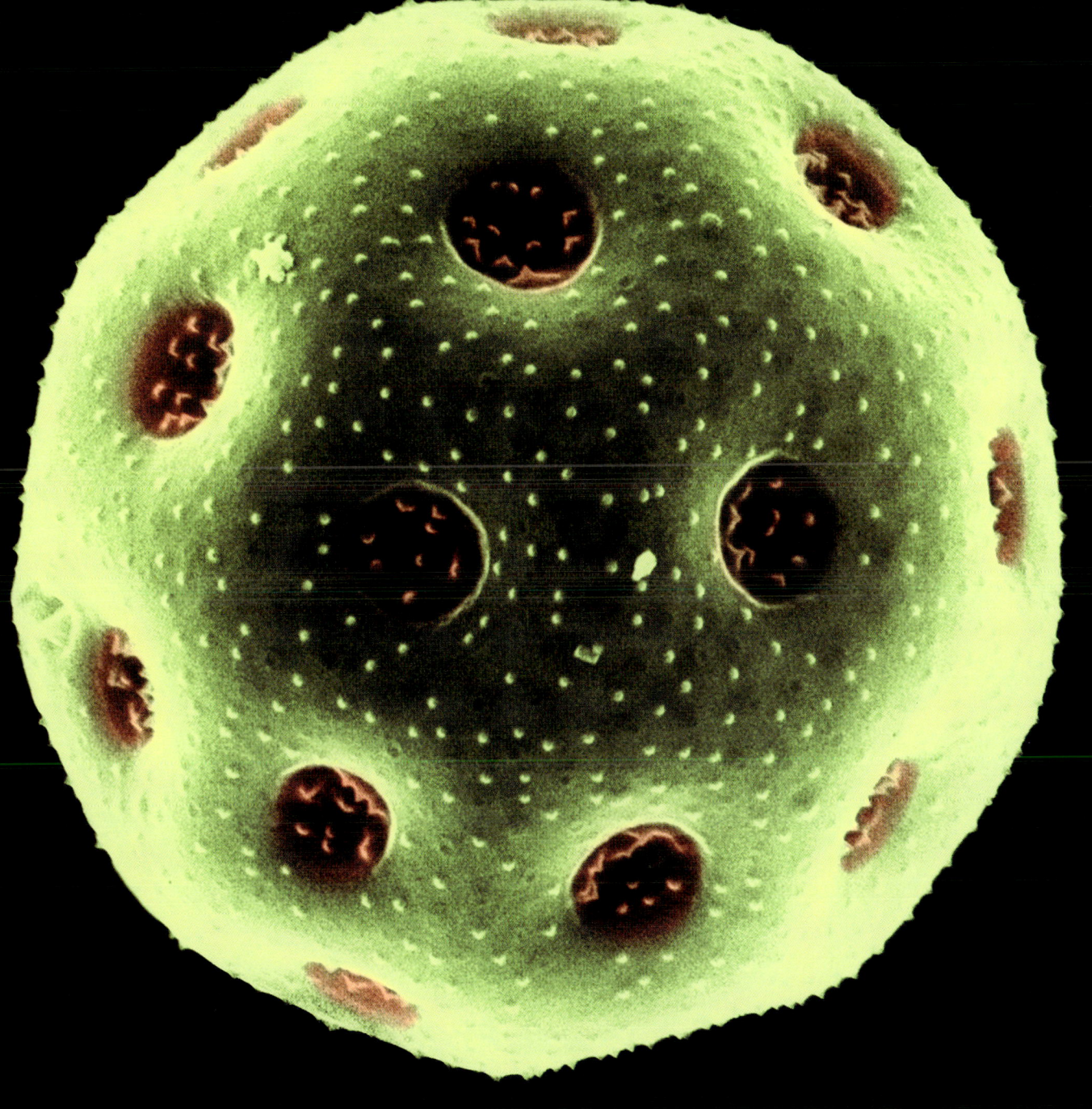

Silene nutans – Nottingham Catchfly (Caryophllaceae) – multi-porate
pollen; each pore is an aperture through which the developing
pollen tube is potentially able to germinate [SEMx 1500]

opposite: *Silene nutans* – Nottingham Catchfly (Caryophllaceae) –
note the stamens which extend well beyond the petals

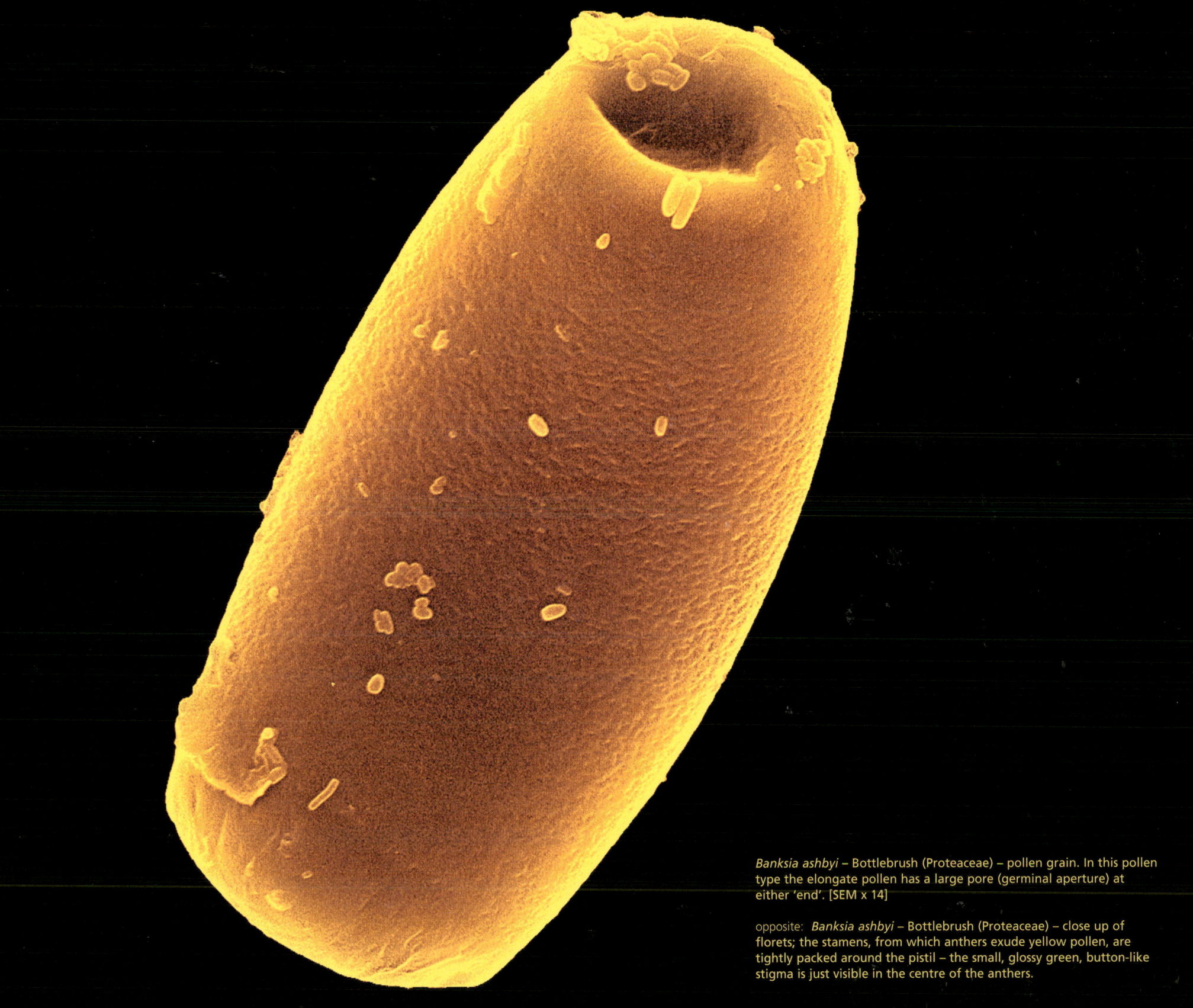

Banksia ashbyi – Bottlebrush (Proteaceae) – pollen grain. In this pollen type the elongate pollen has a large pore (germinal aperture) at either 'end'. [SEM x 14]

opposite: *Banksia ashbyi* – Bottlebrush (Proteaceae) – close up of florets; the stamens, from which anthers exude yellow pollen, are tightly packed around the pistil – the small, glossy green, button-like stigma is just visible in the centre of the anthers.

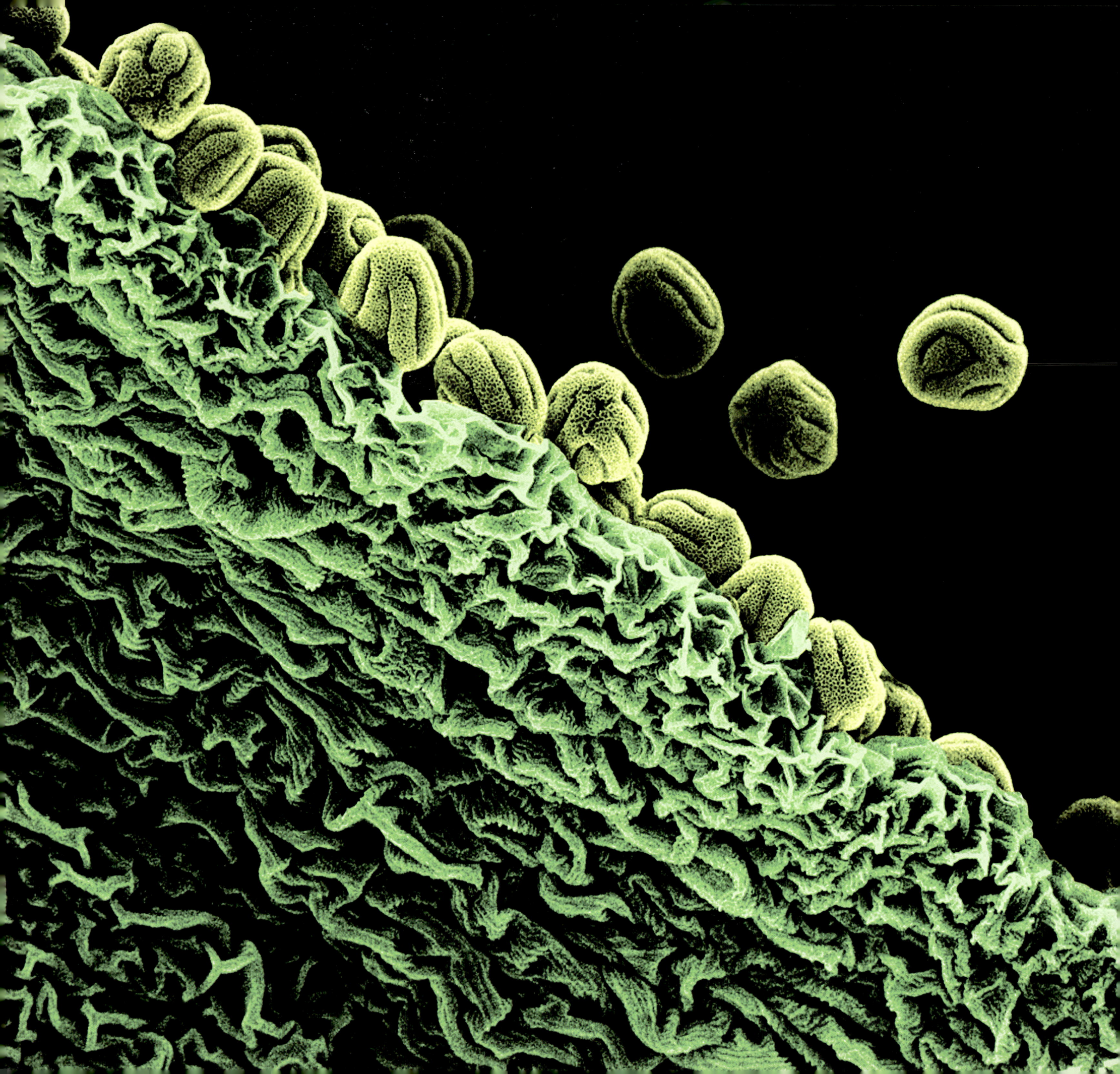

and inexplicable state of mind. Thus the sight of a flower reveals, it is true, the presence of this well-defined part of the plant, the sight of the flower provokes in the mind much more significant reactions, because the flower expresses an obscure vegetal resolution. What the configuration and colour of the corolla reveal, what the dirty traces of pollen or the freshness of the pistil betray doubtless cannot be adequately expressed by language; it is, however, useless to ignore (as is generally done) this inexpressible real presence and to reject as puerile absurdities certain attempts at symbolic interpretation.

Georges Bataille

The Language of Flowers, Documents 3, 1929

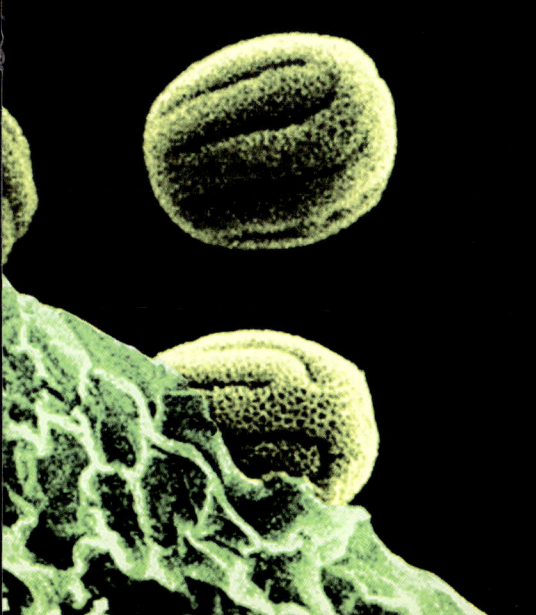

Primula veris – Cowslip (Primulaceae) –
pollen grains along the open edge of an anther theca [SEM x 3000]

195

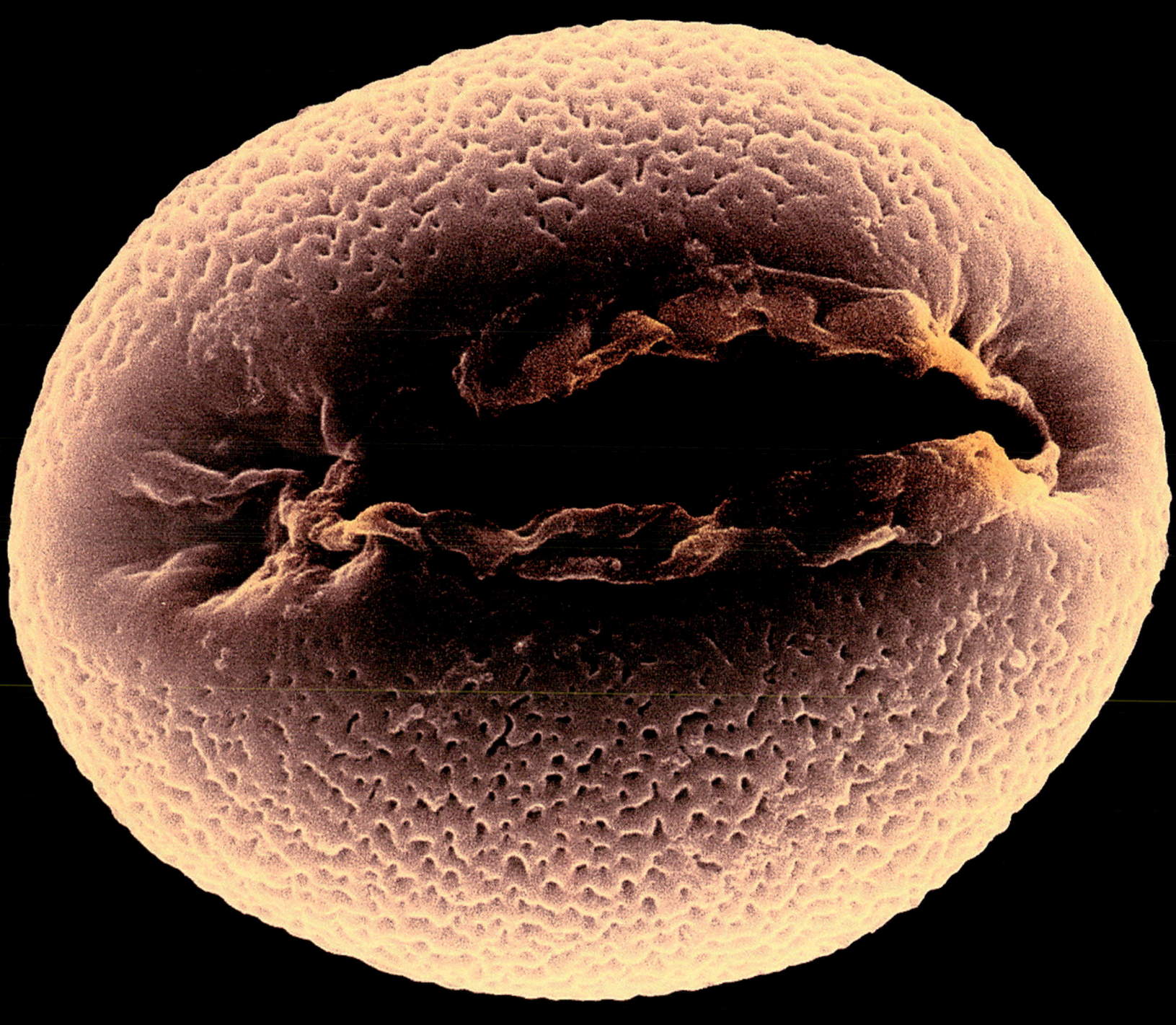

Phoenix canariensis – Canary Islands Palm (Arecaceae) –
pollen grain [SEM x 5000 – acetolysed]

opposite: *Verbascum nigrum* – Dark Mullein (Scrophulariaceae) –
pollen grain [SEM x 3000 – acetolysed]

Bid the closed *petals* from nocturnal cold

The virgin *Style* in silken curtains fold,

Shake into viewless air the morning dews,

And wave in light their iridescent hues;

While from on high the bursting *Anthers* trust

To mild breezes their prolific dust;

Or bend in rapture o'er the central Fair,

Love out their hour, and leave their lives in air.

Love out their hour. The vegetable passion of love is agreeably seen in the flower of the paranassia, in which the males alternately approach and recede from the female, and in the flowers of the nigella, or devil in the bush, in which the tall females bend down to dwarf their husbands. But I was this morning surprised to observe, amongst Sir Brooke Boothby's valuable collection of plants at Ashbourn, the manifest adultery of several females of the plant Collinsonia, who had bent themselves into contact with the males of other flowers of the same plant in their vicinity, neglectful of their own.

Erasmus Darwin

The Botanic Garden, 1791

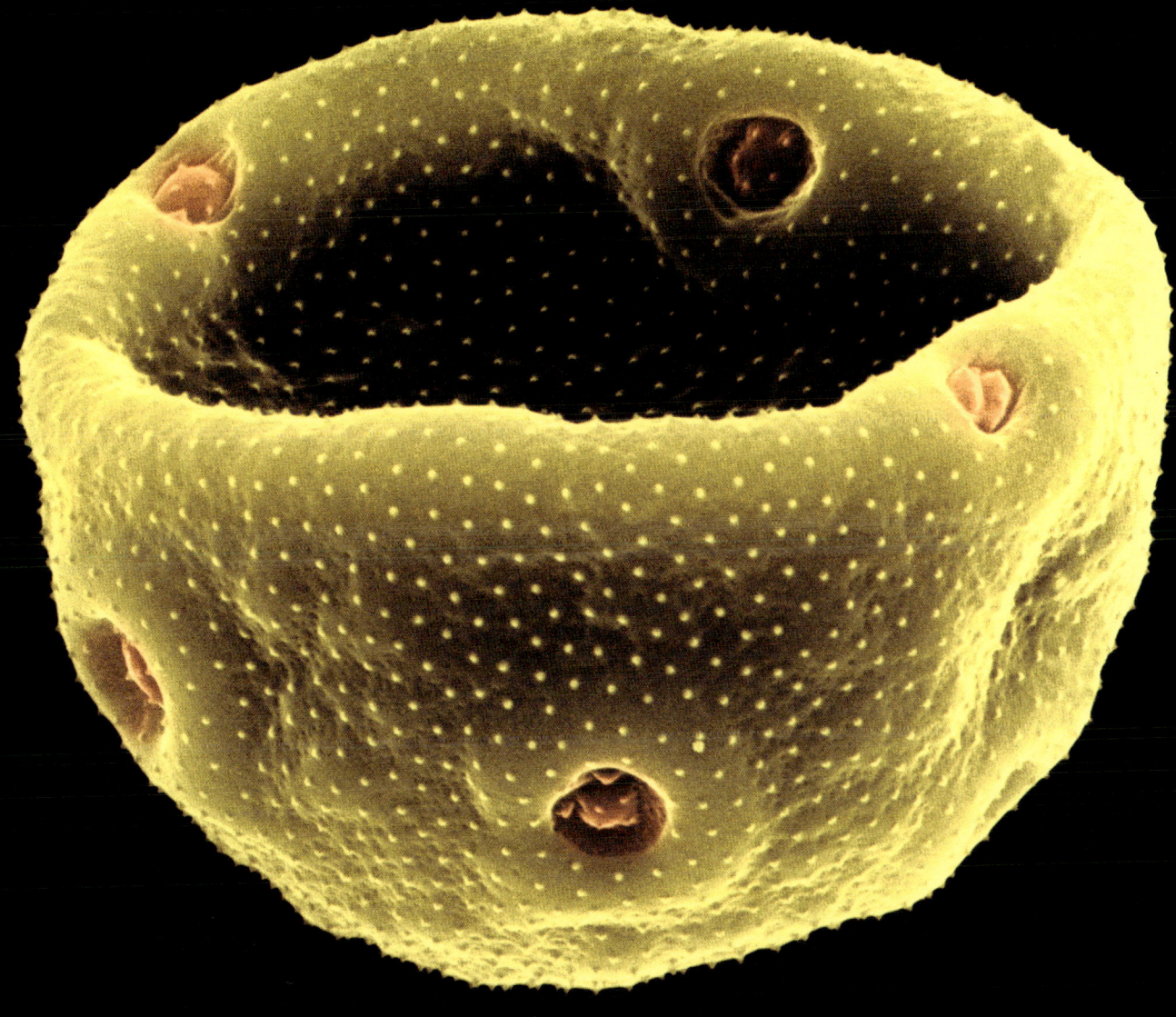

Plantago lanceolata – Ribwort Plantain (Plantaginaceae) –
multiporate pollen grain, which is completely dehydrated and folded
inwards like a deflated ball. [SEM x 3000]

opposite: *Plantago lanceolata* – Ribwort Plantain (Plantaginaceae) –
inflorescence viewed from above; note the slender filaments of the
stamens which, like the stamens of grass flowers, allow the anthers to
catch the wind (cf. grass flowers).

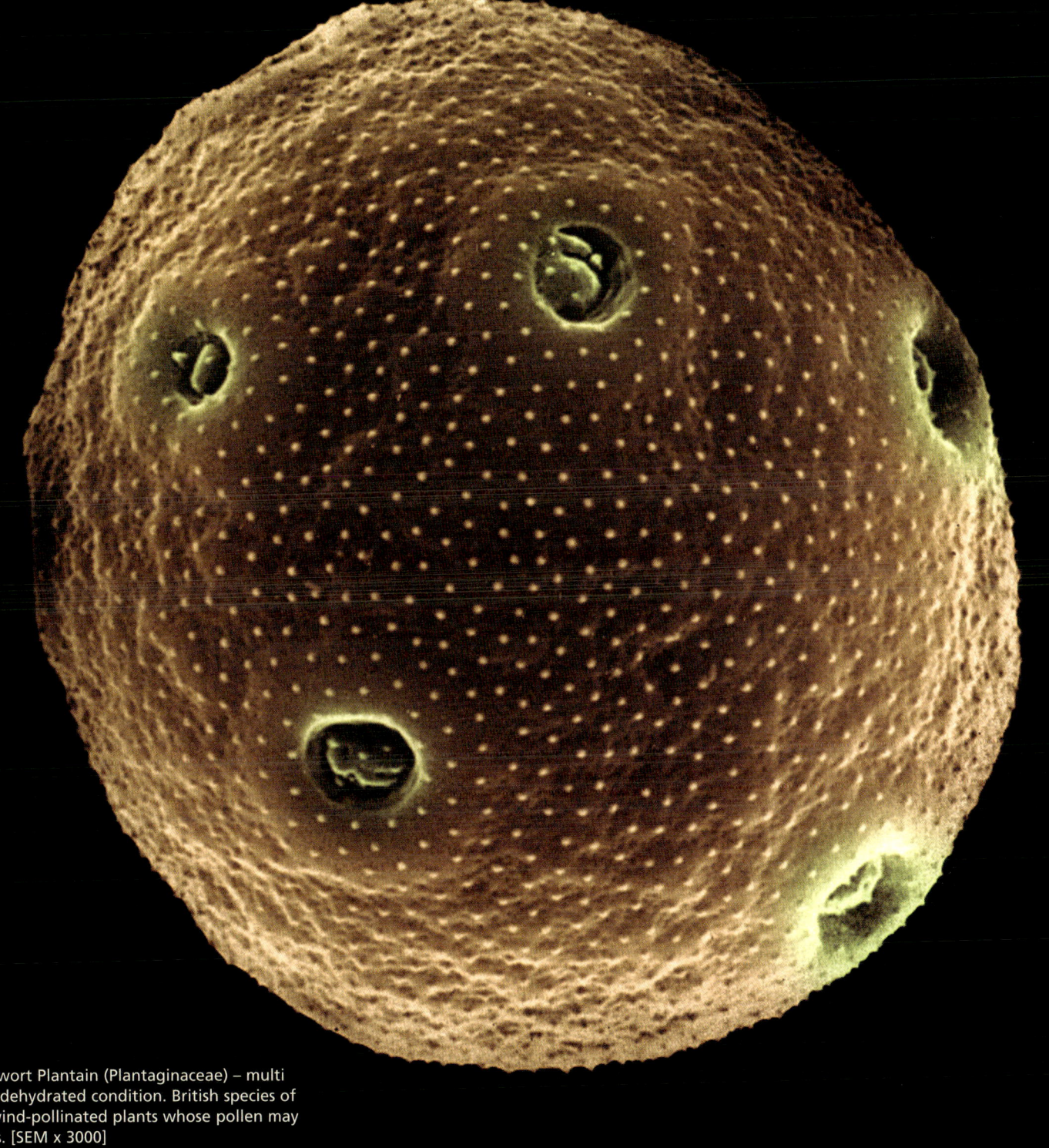

Plantago lanceolata – Ribwort Plantain (Plantaginaceae) – multi porate pollen grain, semi dehydrated condition. British species of Plantain are among the wind-pollinated plants whose pollen may cause hay fever symptoms. [SEM x 3000]

opposite: *Plantago lanceolata* – Ribwort Plantain (Plantaginaceae) – inflorescence viewed from side.

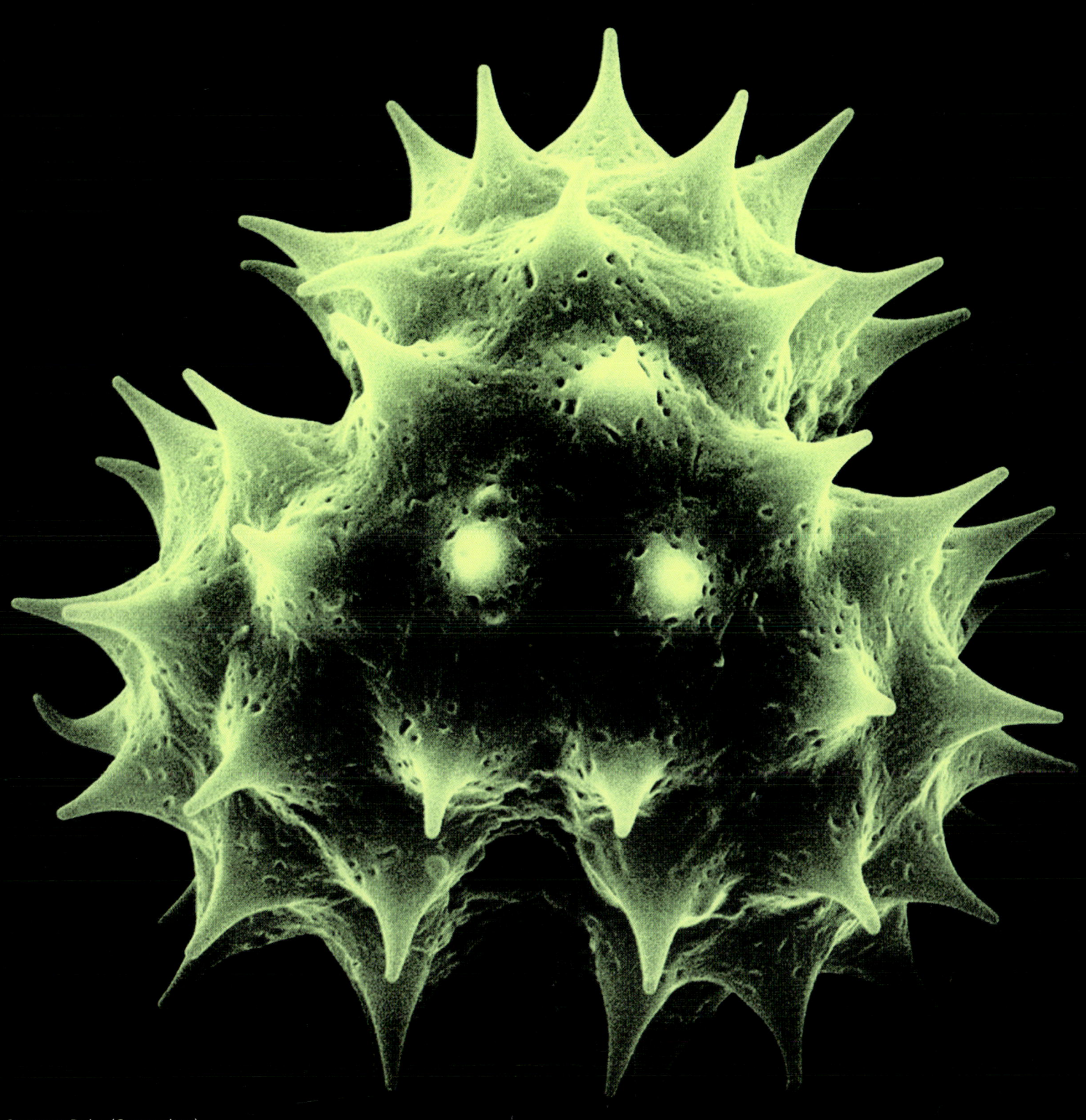

Bellis perennis – Common Daisy (Compositae) –
pollen grain [SEM x 3000]

opposite: *Bellis perennis* – Common Daisy (Compositae) –
a composite flower

Primula veris – Cowslip (Primulaceae) – this pollen grain has not developed normally, and lacks a regular aperture arrangement. (SEM x 3000)

opposite: *Primula veris* – Cowslip (Primulaceae) – inflorescence, 'thrum' (anthers emerging before pistil) arrangement (cf. Polyanthus).

overleaf left: *Euphorbia amygdaloides* – Wood Spurge (Euphorbiaceae) – inflorescences; the large outer petal-like, kidney-shaped structures are 'bracts' or 'prophylls'; they surround an open cyathium flanked by a pair of young, unopened bracts. The cyathium consists of an outer cup-shaped structure bearing four horseshoe-shaped glands on the rim; within the cup is a group of condensed flowers, comprising a ring of male flowers each consisting of a single stamen; in the centre is a female flower comprising a stalked ovary with three branched stigmas.

overleaf right: *Euphorbia amygdaloides* – Wood Spurge (Euphorbiaceae) – three pollen grains [SEM x 1000]

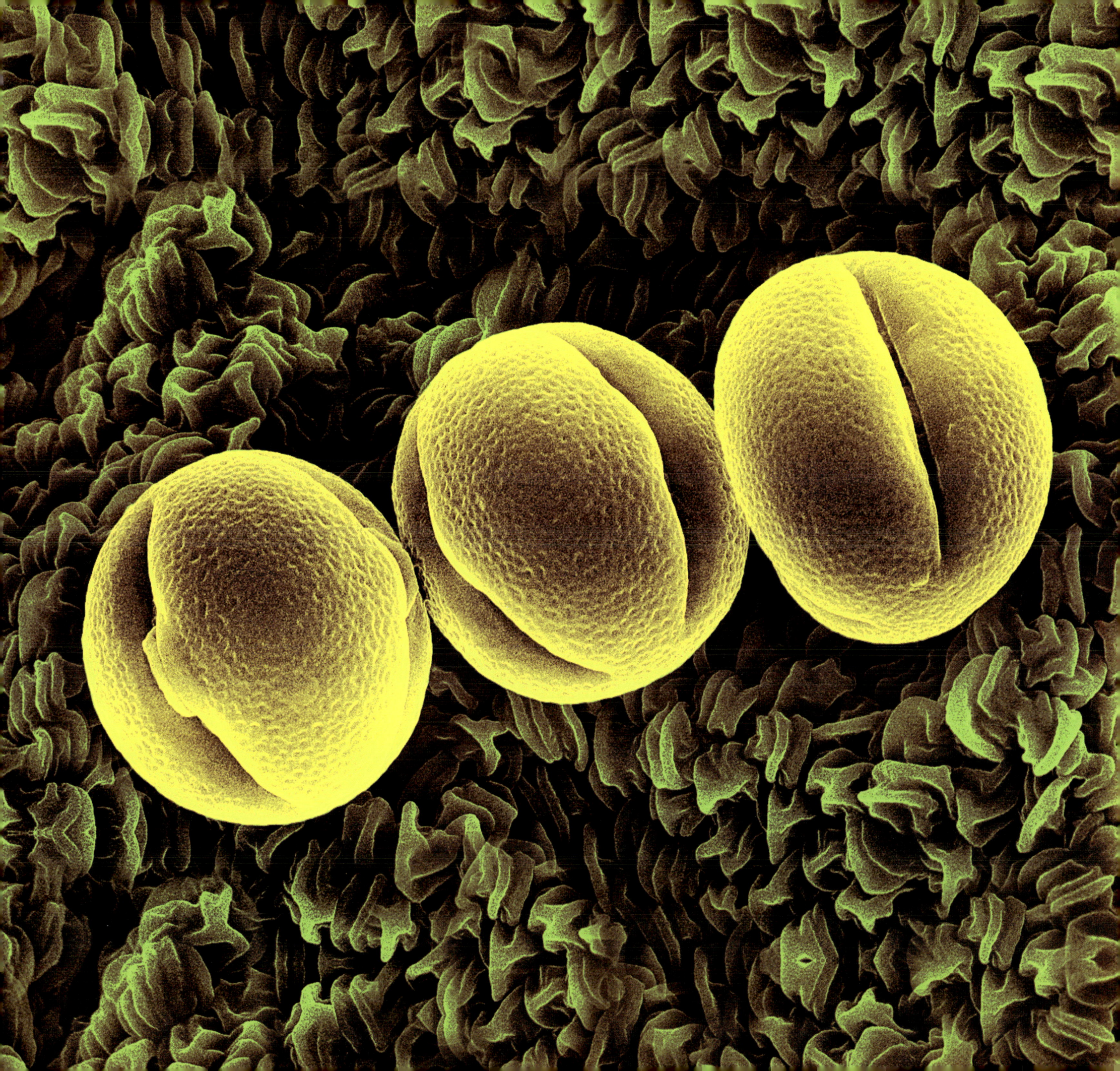

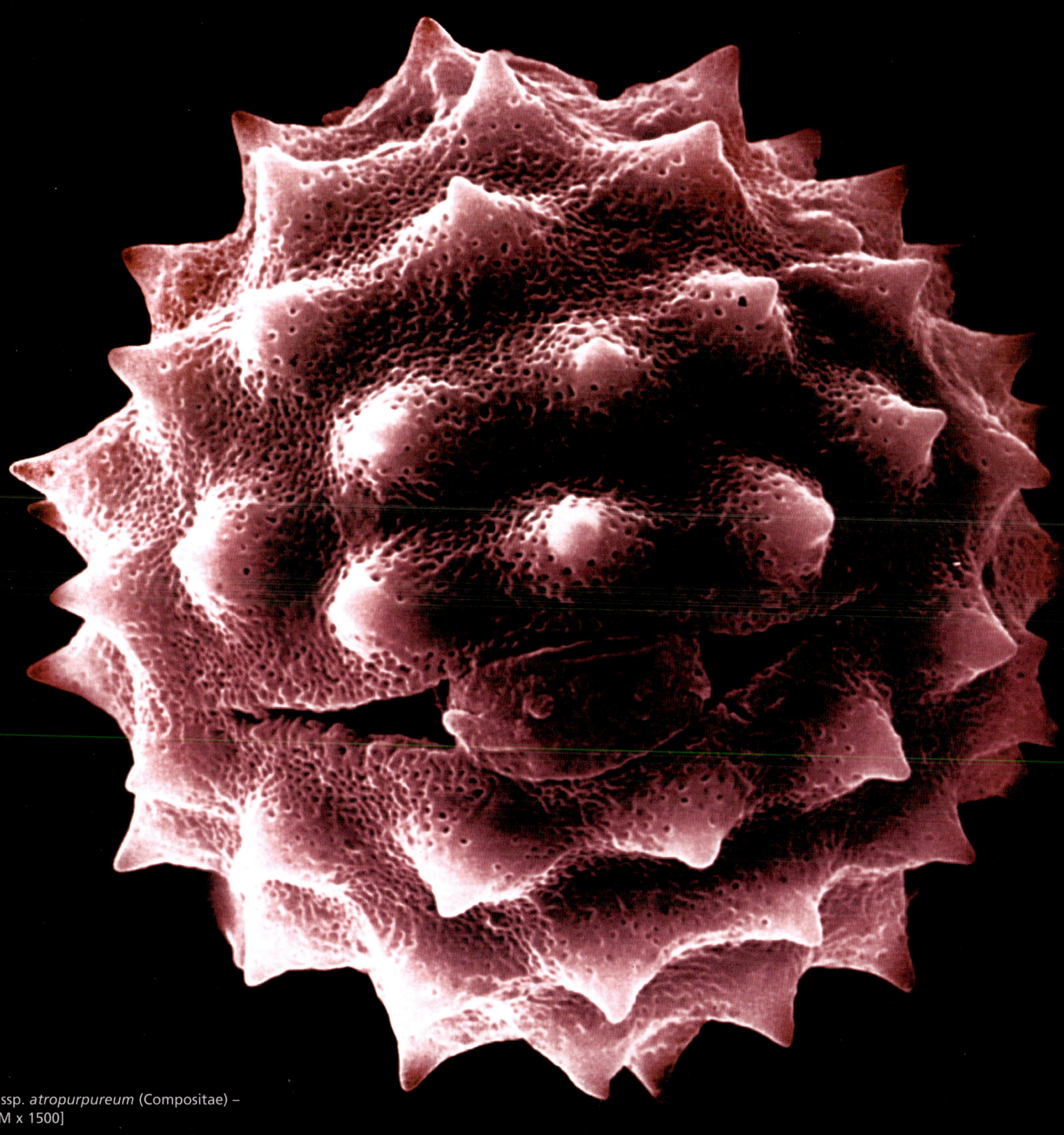

Cirsium rivulare ssp. *atropurpureum* (Compositae) –
pollen grain [SEM x 1500]

opposite: *Cirsium rivulare* ssp. *atropurpureum* (Compositae) –
a composite flower

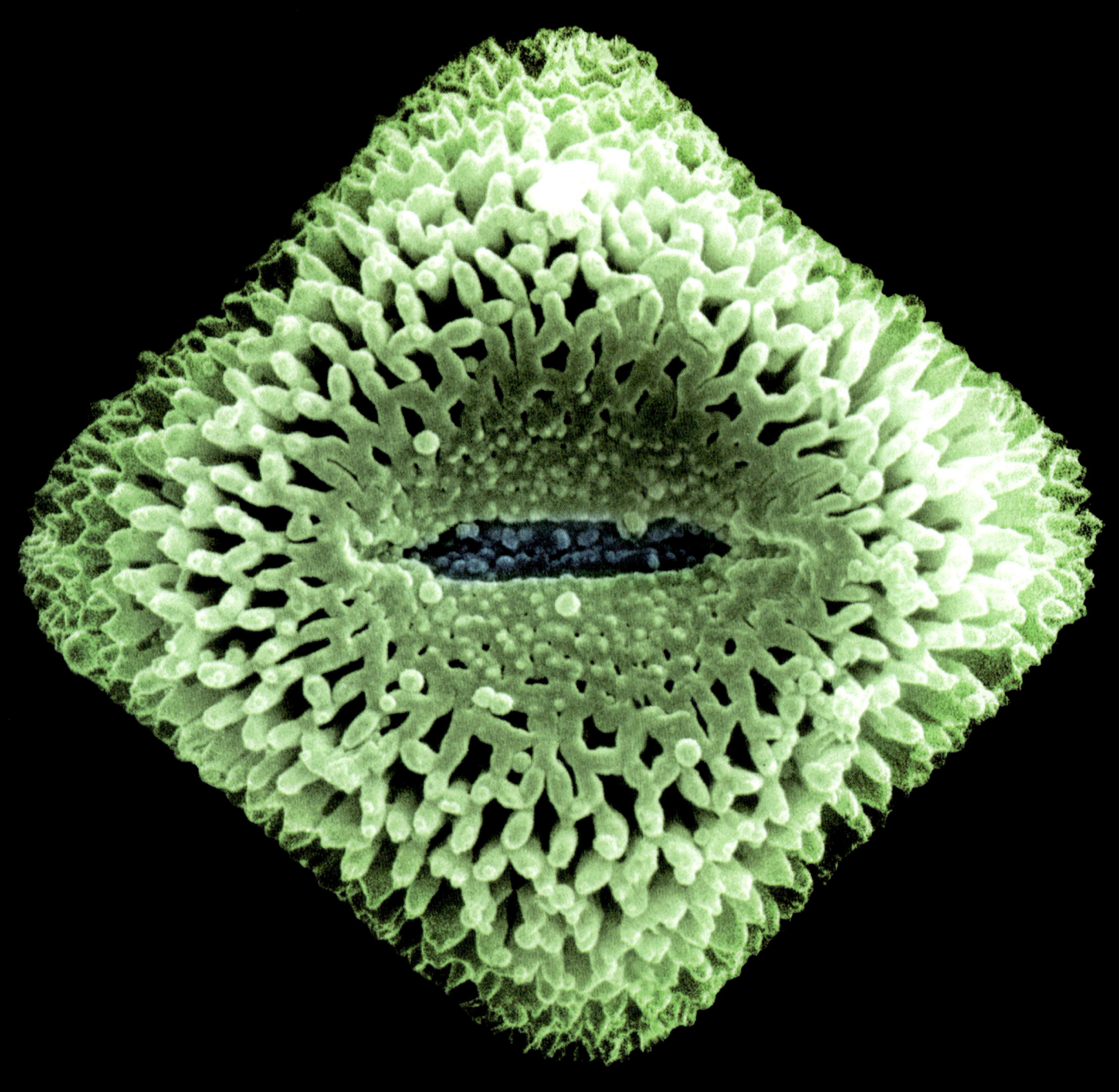

Basella alba – Malabar Spinach (Basellaceae) –
pollen grain [SEM x 1600 – acetolysed]

opposite: *Basella rubra* – Malabar Nightshade (Basellaceae) –
pollen grain [SEM x 1800 – acetolysed]

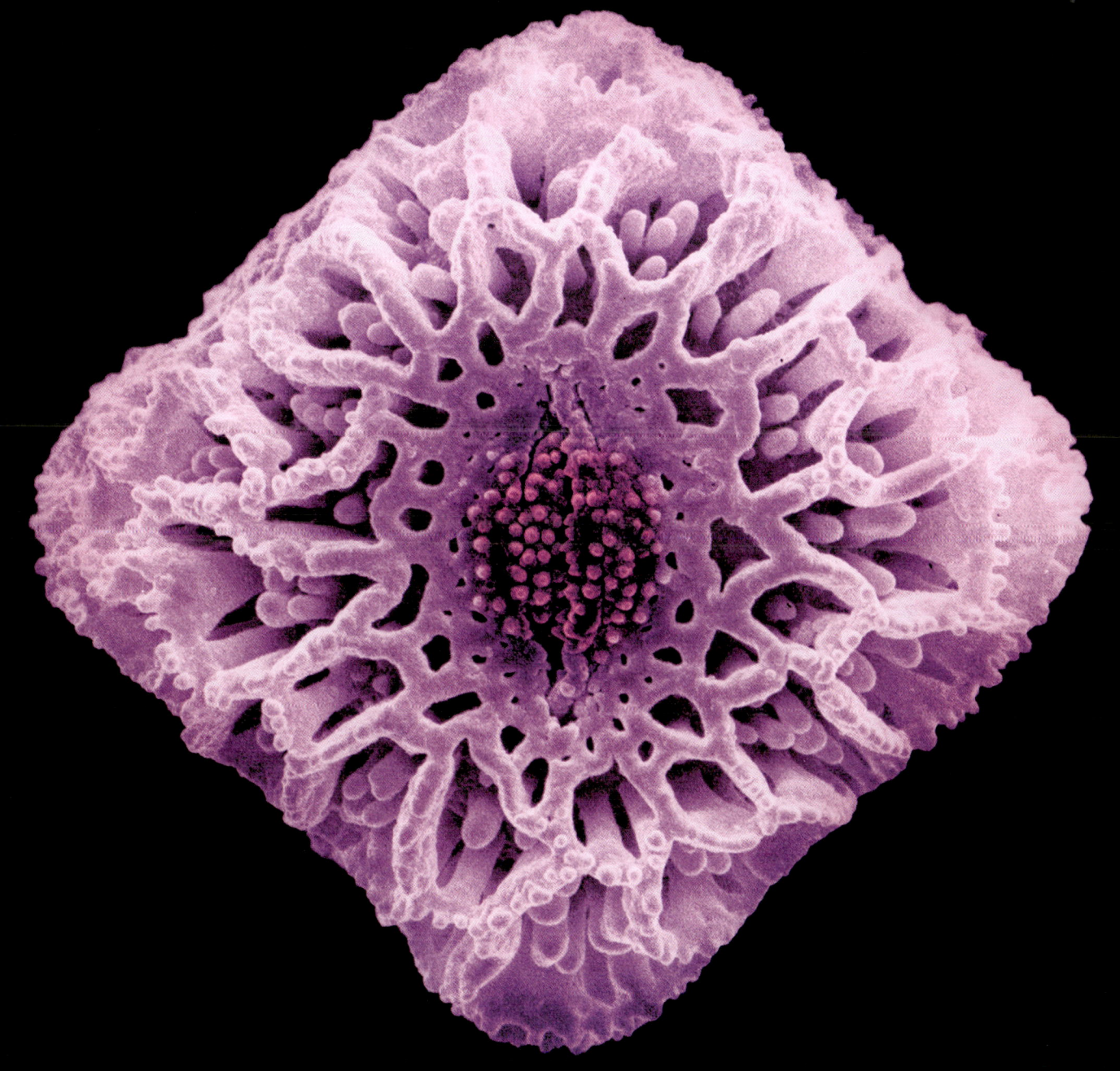

overleaf left: *Aesculus Hippocastanum* – Horse-chestnut
(Hippocastanaceae) – close up of a floret from a large inflorescence

overleaf right: *Aesculus Hippocastanum* – Horse-chestnut
(Hippocastanaceae) – pollen grain [SEM x 3000]

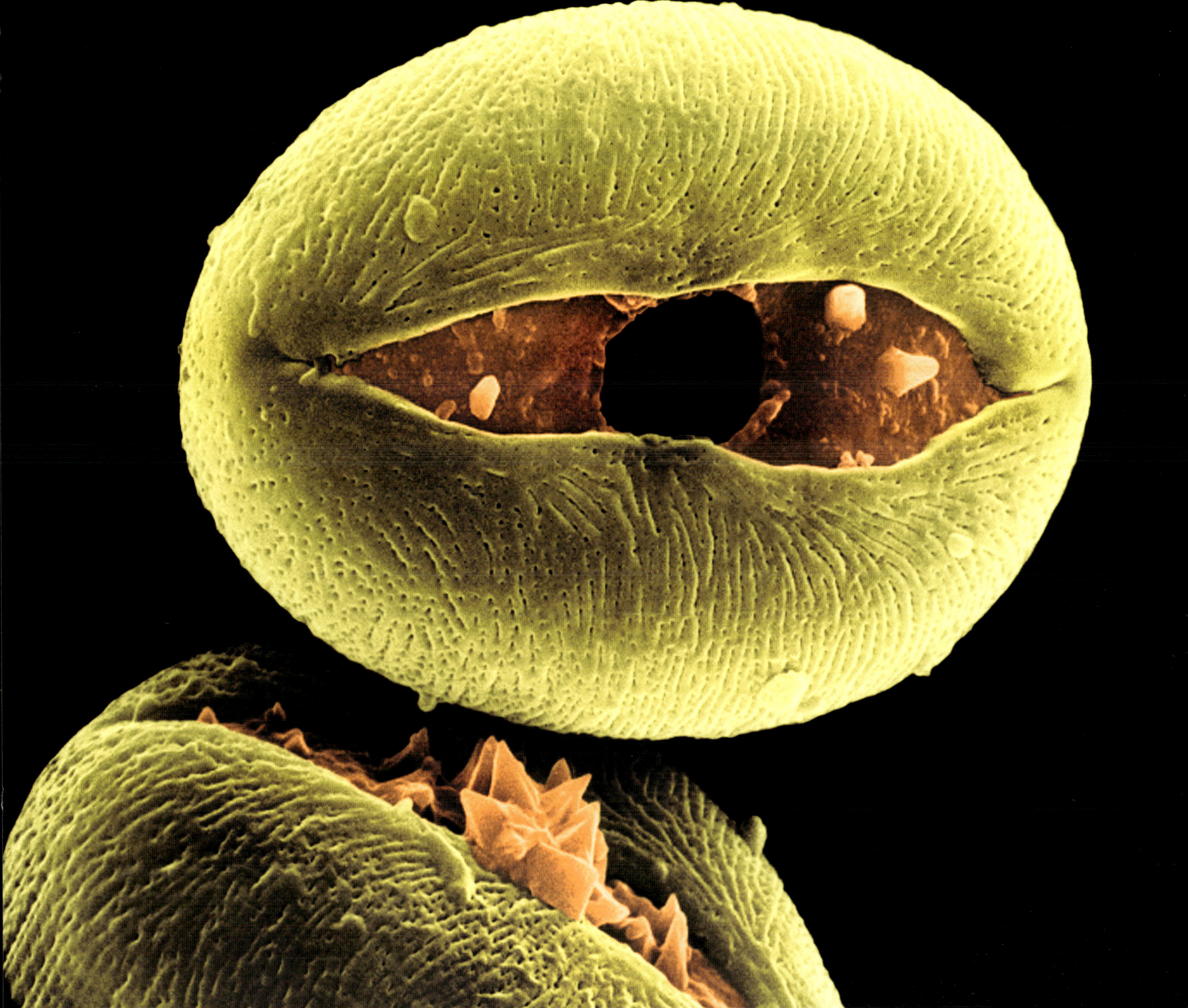

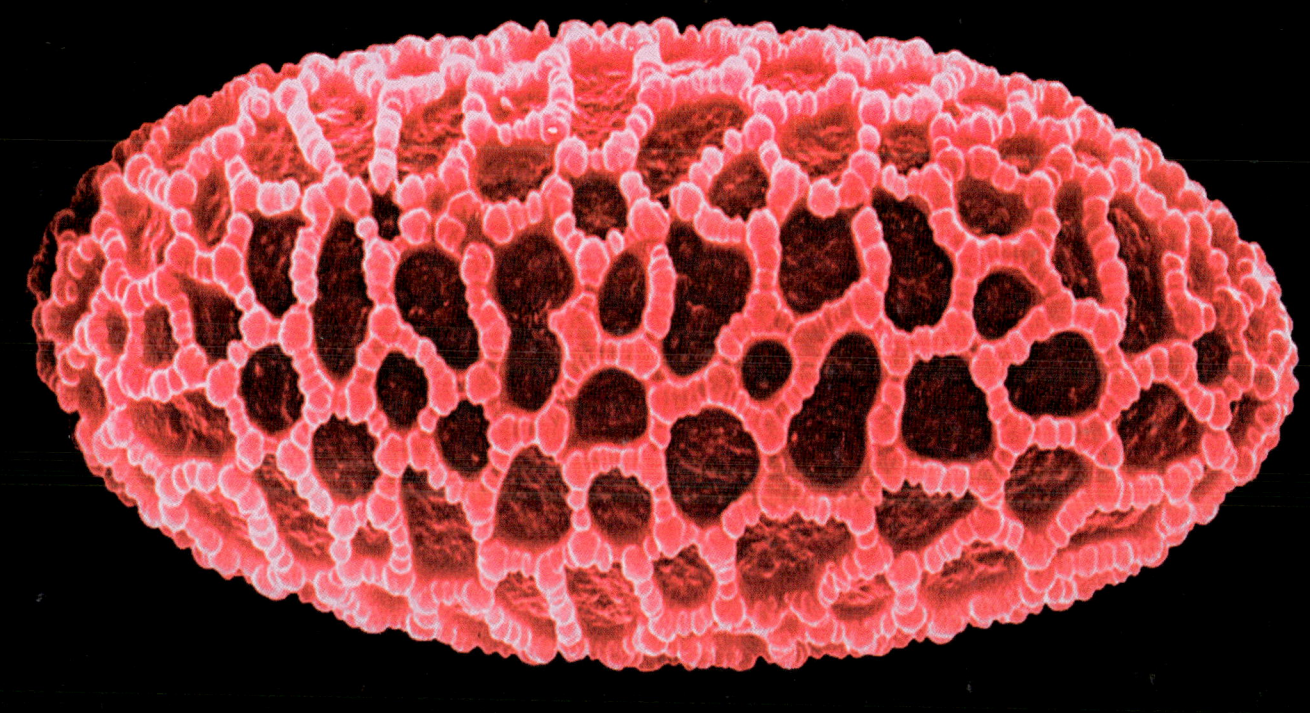

Nature does not create works of art.

It is we, and the faculty of interpretation peculiar to the human mind, that see art.

Man Ray

Nerine bowdeni (Amaryllidaceae) – pollen grain
[SEM x 1000]

opposite: *Nerine bowdeni* (Amaryllidaceae) – an inflorescence

overleaf left: *Asphodelus microcarpus* – Common Asphodel [in Greece]
(Asphodelaceae) – single flower from a spikelet of flowers

overleaf right: *Narthecium ossifragum* – Bog Asphodel (Nartheciaceae)
– pollen grain, semi dehydrated condition [SEM x 5000]

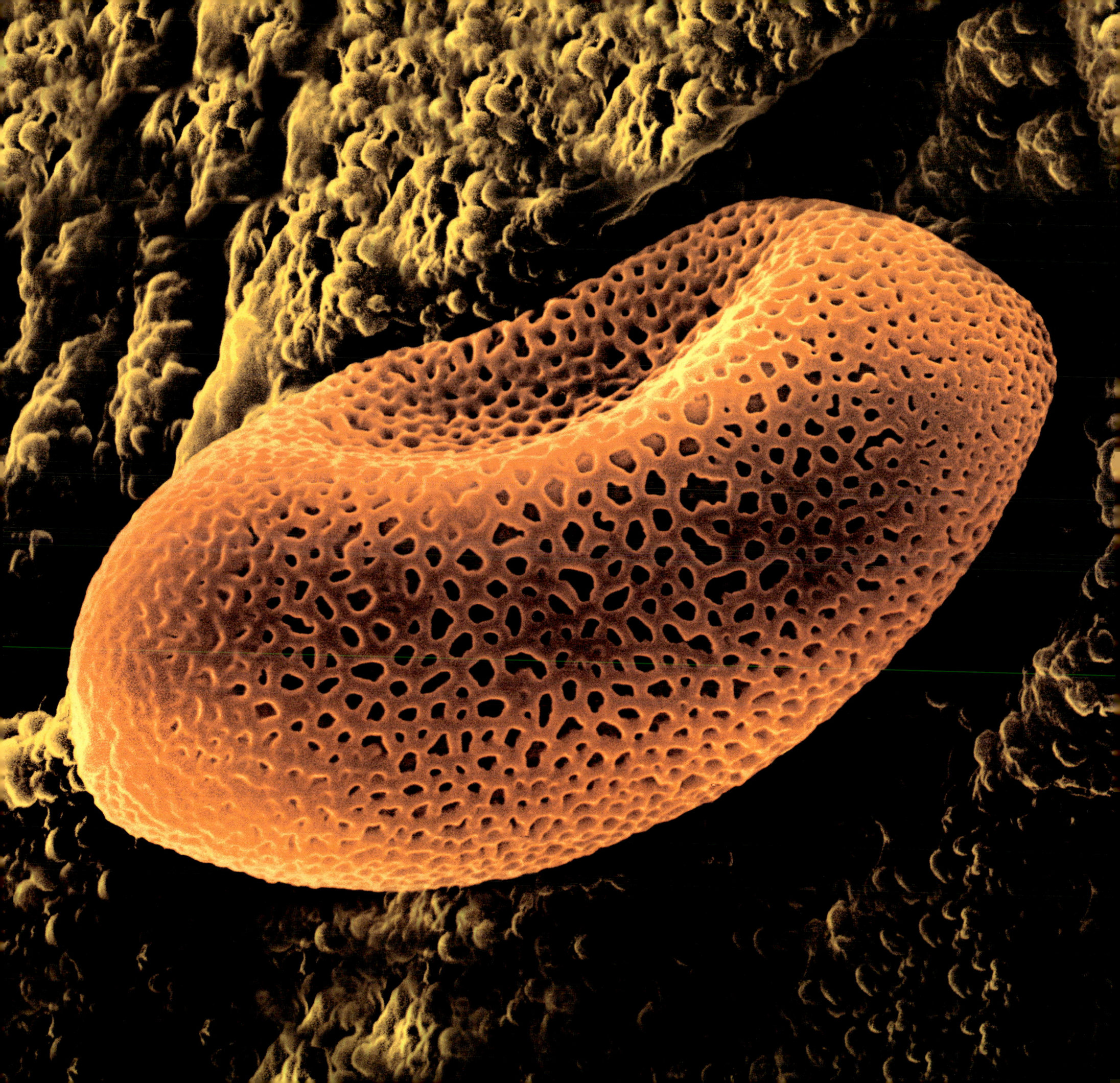

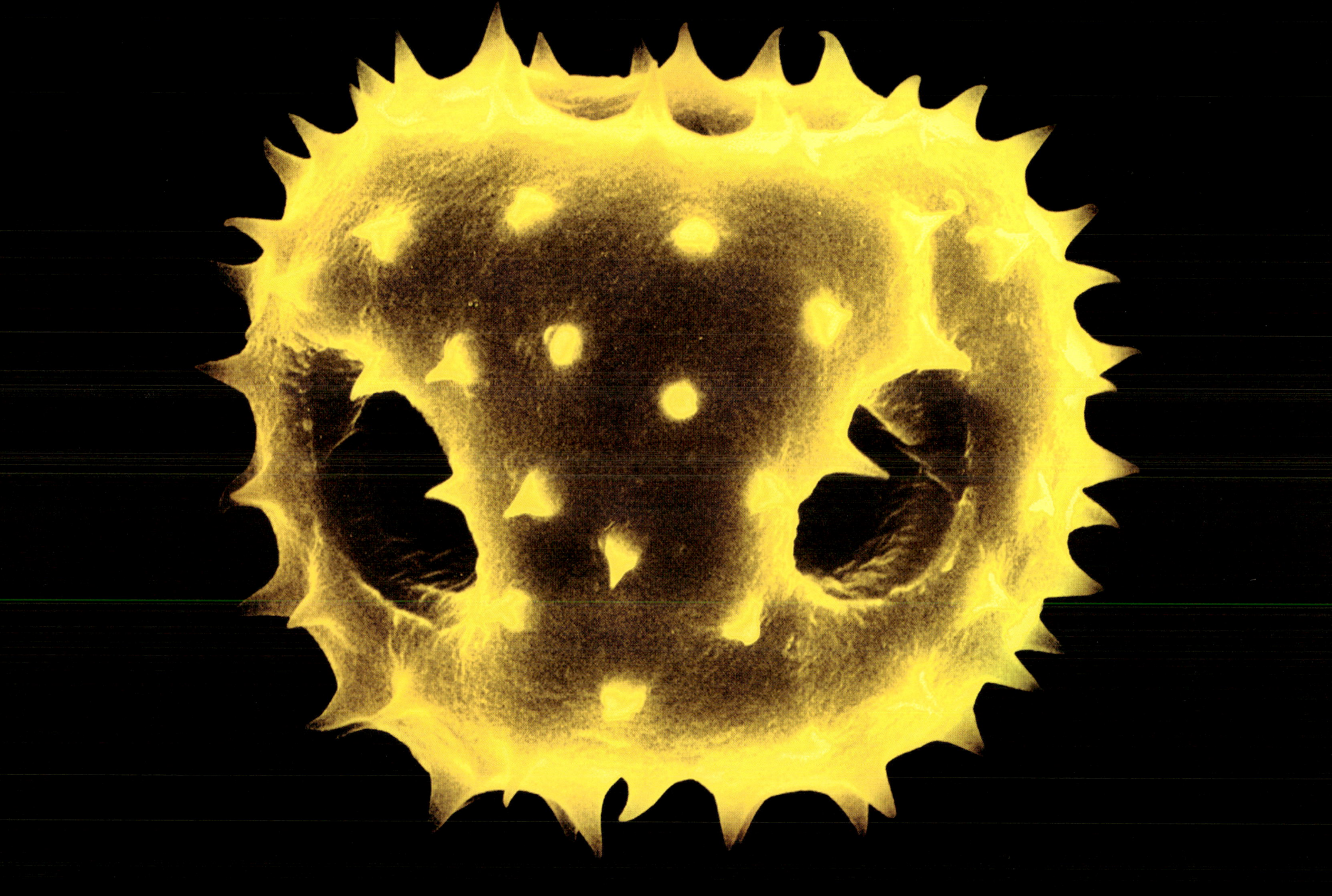

Tussilago farfara – Coltsfoot (Compositae) – pollen grain,
semi hydrated condition [SEM x 1500]

opposite: *Tussilago farfara* – Coltsfoot (Compositae) –
a composite flower

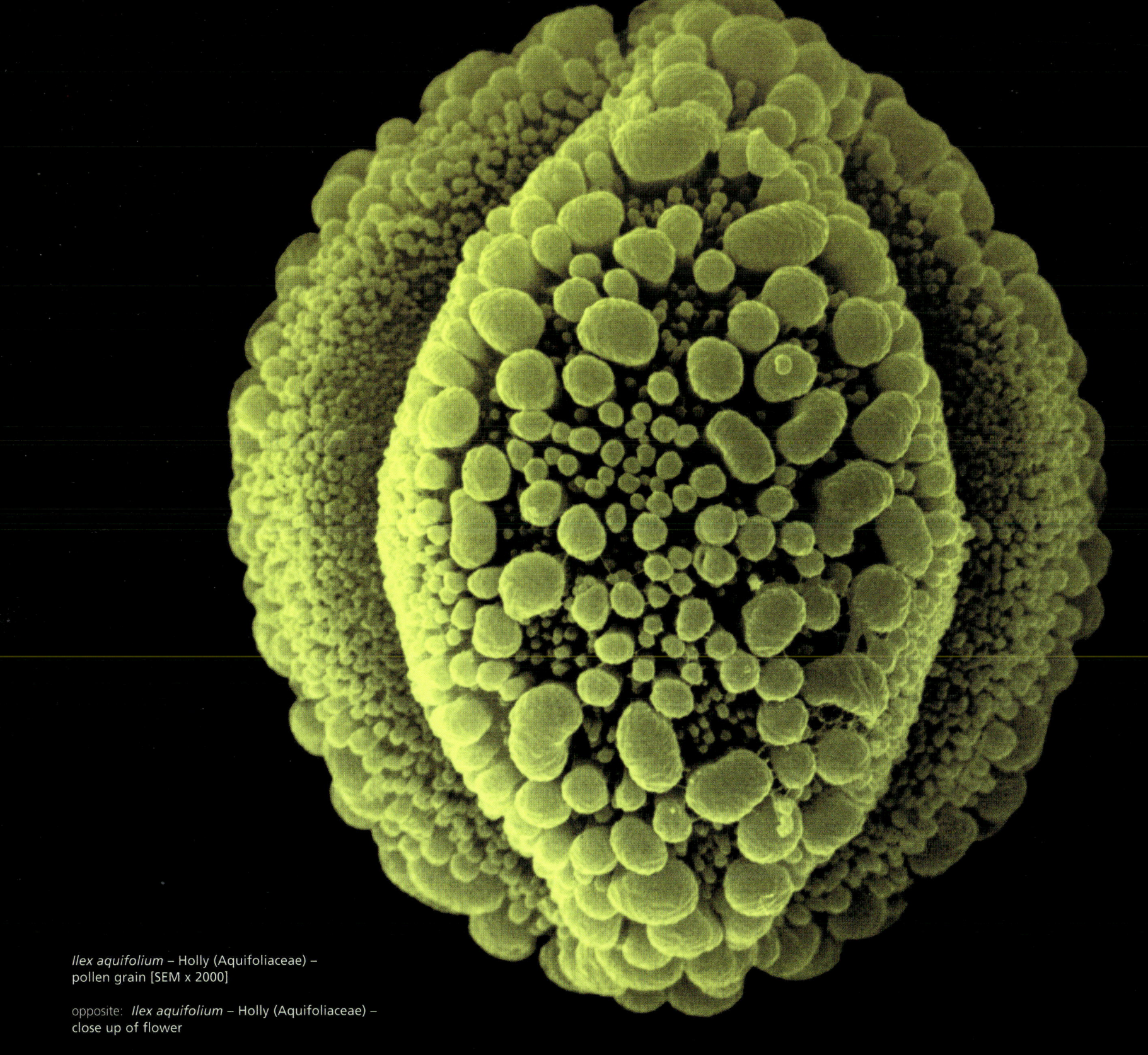

Ilex aquifolium – Holly (Aquifoliaceae) –
pollen grain [SEM x 2000]

opposite: *Ilex aquifolium* – Holly (Aquifoliaceae) –
close up of flower

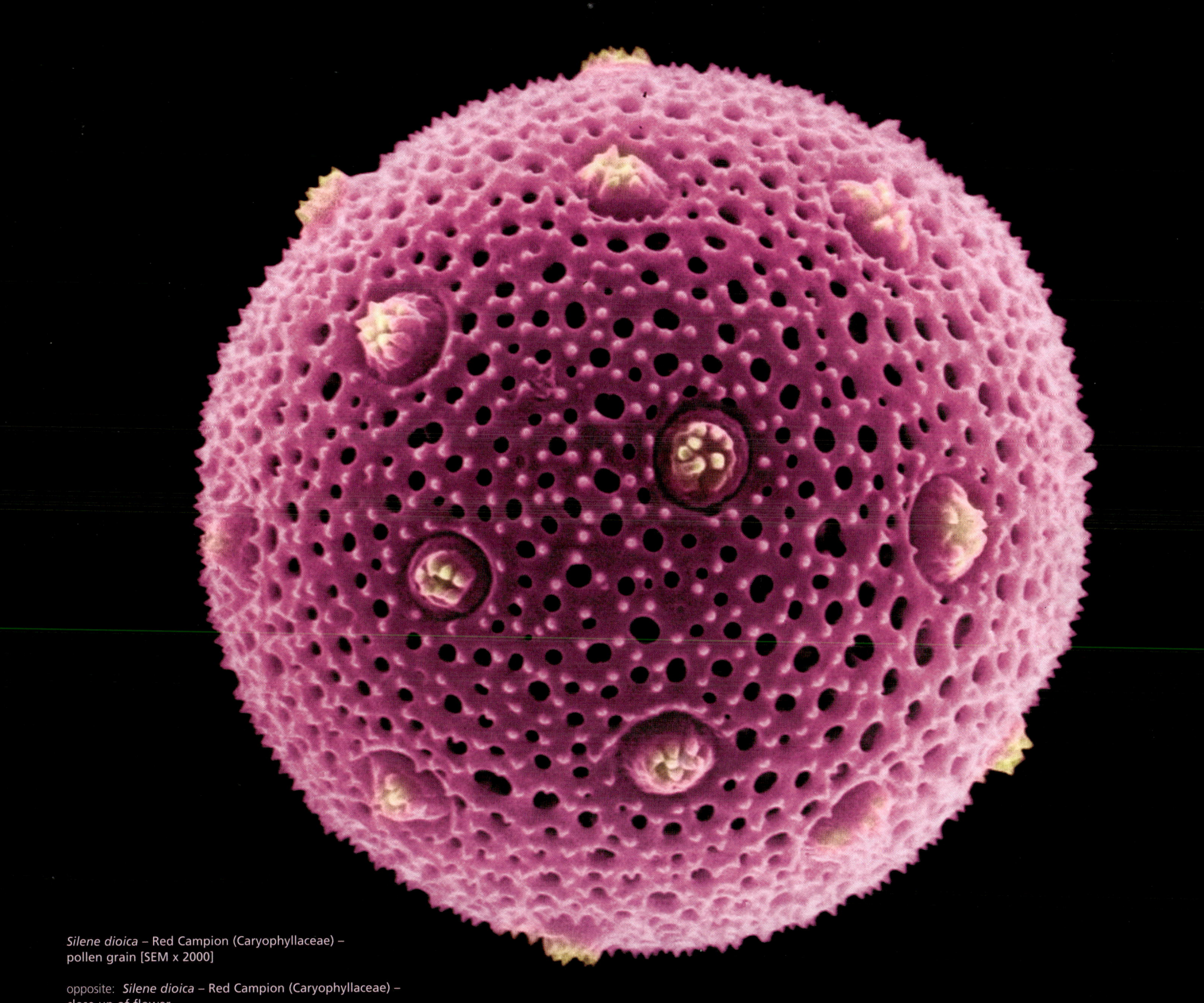

Silene dioica – Red Campion (Caryophyllaceae) –
pollen grain [SEM x 2000]

opposite: *Silene dioica* – Red Campion (Caryophyllaceae) –
close up of flower

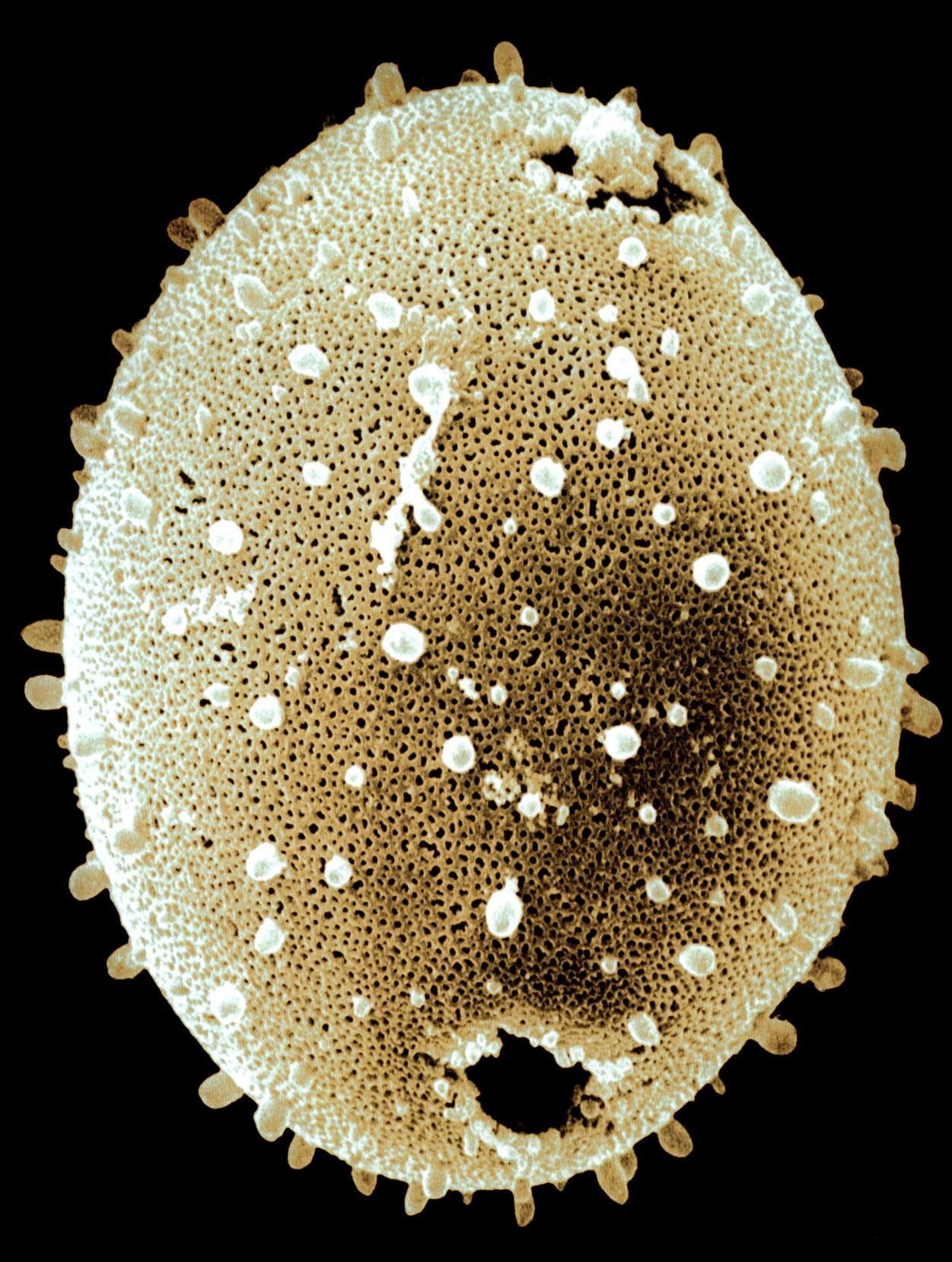

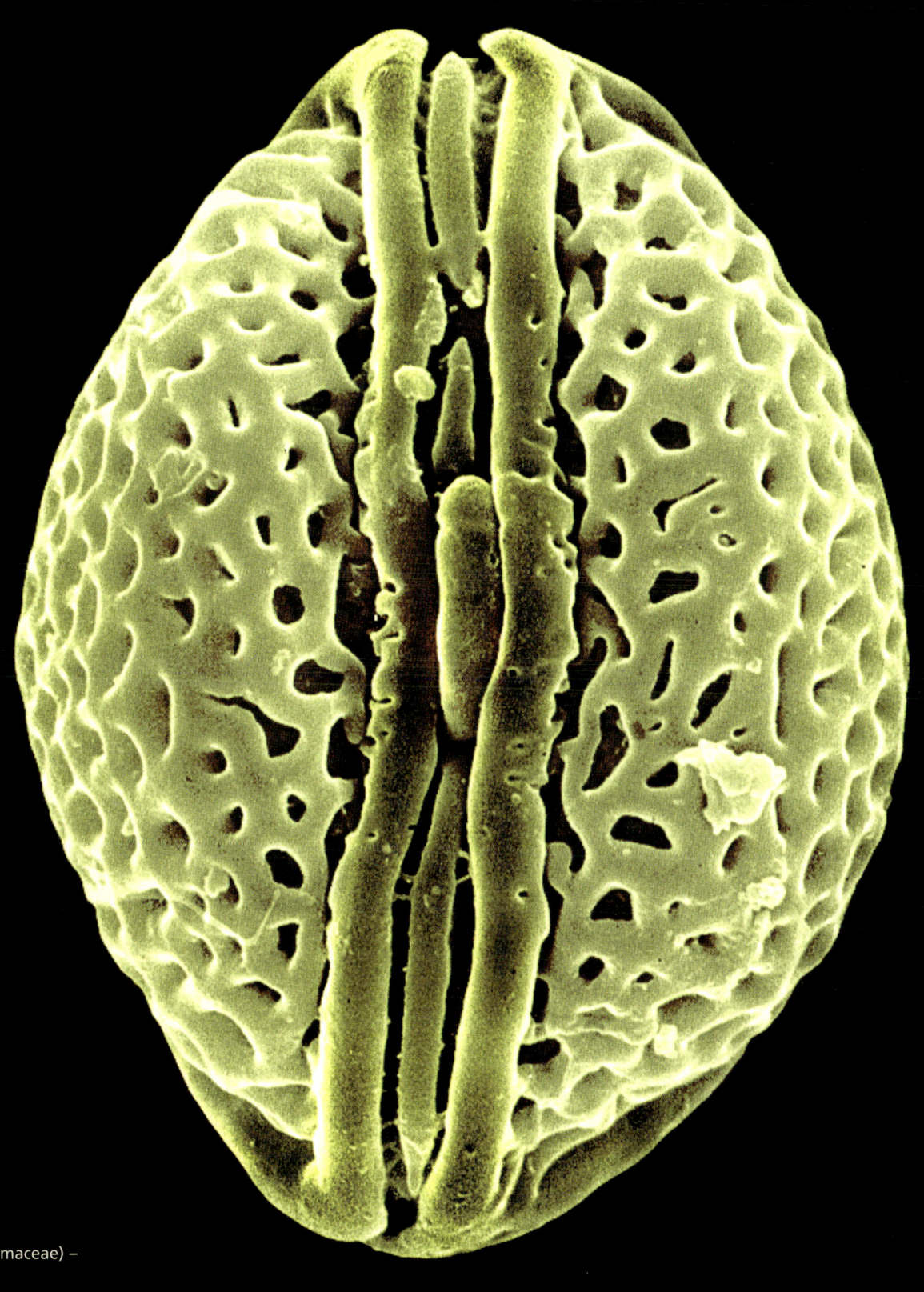

Tinospora smilacina (Menispermaceae) –
pollen grain [SEM x 3200]

opposite: *Adansonia digitata* – Baobab (Bombacaceae) –
pollen grain [SEM x 1200]

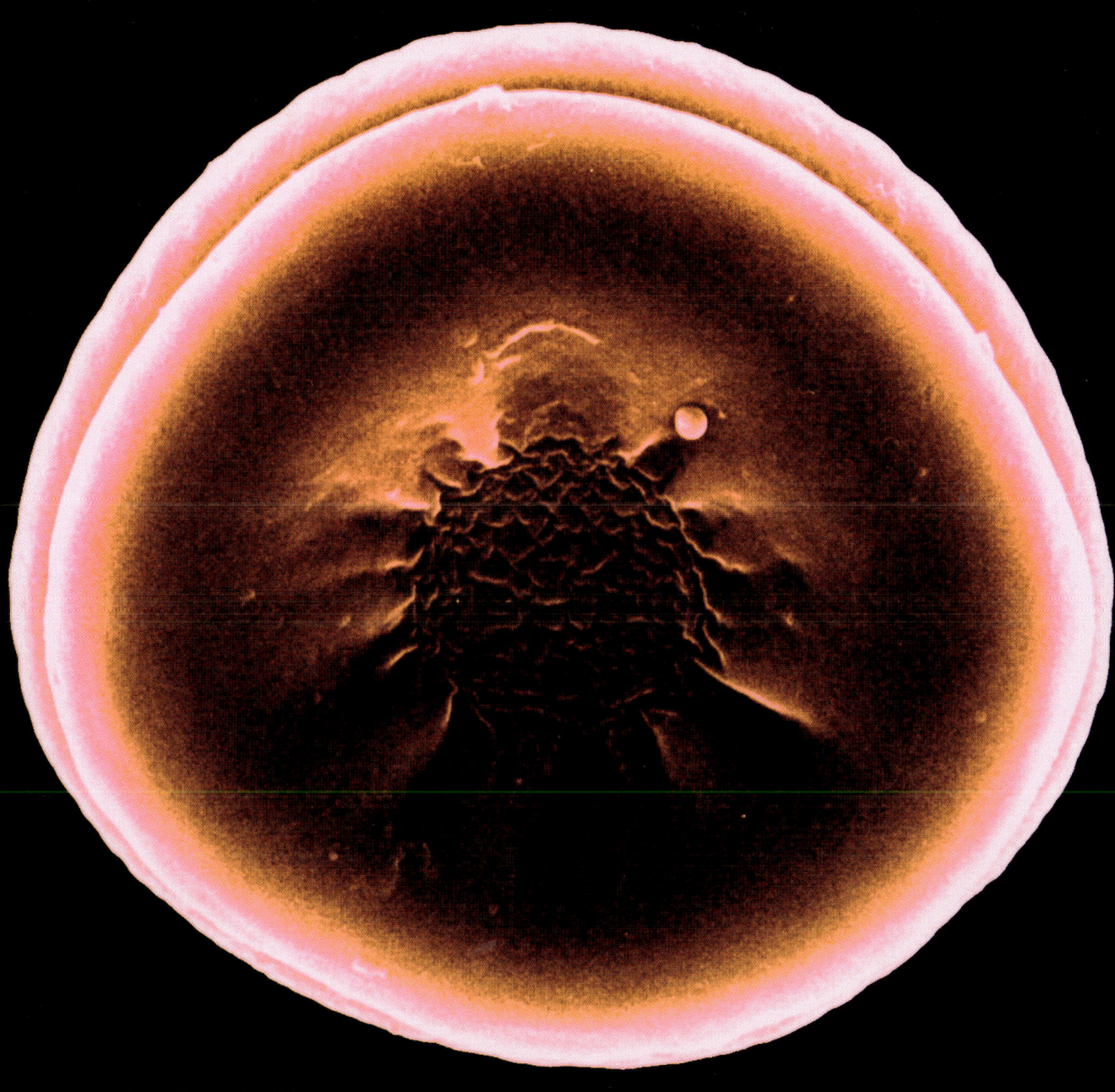

Nymphaea cv. – Water Lily (Nymphaeaceae) – pollen grain, slightly dehydrated. This type of pollen has a generally smooth exine, and a ring-like aperture around the perimeter of the grain, rather like a hamburger bun. The central wrinkled area of the upper surface, although not obvious in the picture, is thinner than the smooth exine, which surrounds it. When the pollen grain dehydrates the thin area contracts. (SEM x 1500)

opposite: *Nymphaea* cv. – Water Lily (Nymphaeaceae) – flower, the crown of small inner petals grade into the numerous central stamens.

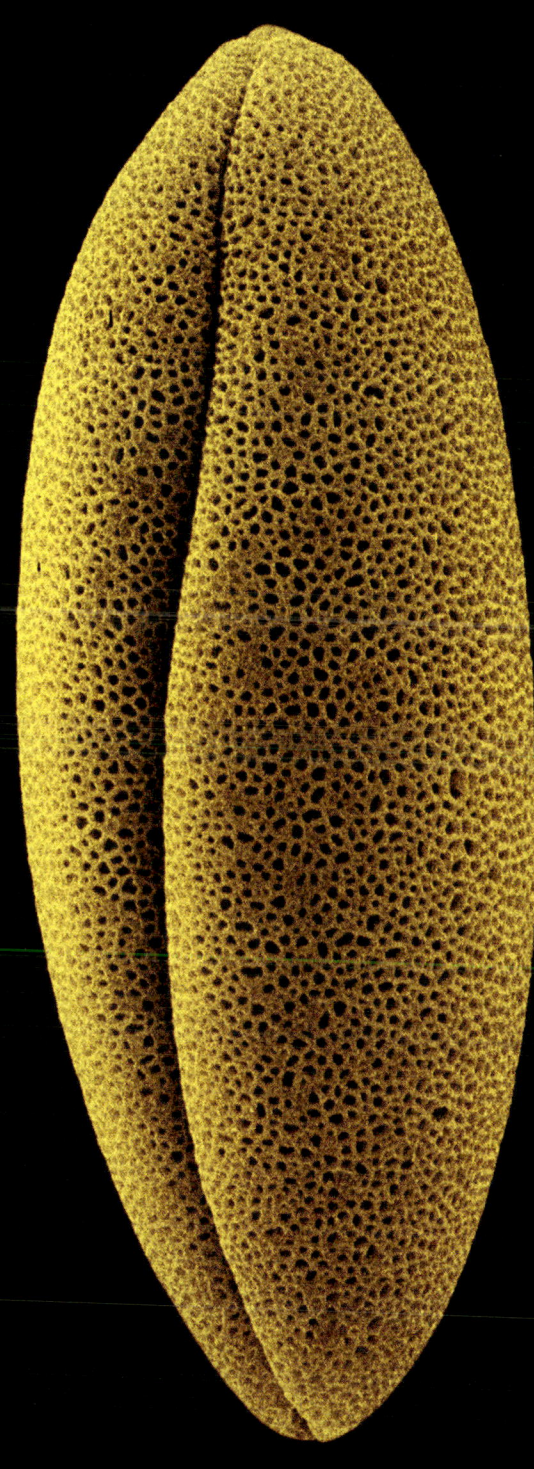

Narcissus cv. – Daffodil (Amaryllidaceae) – pollen grain, dehydrated
condition, the single slit like aperture is infolded.
[SEM x 2000]

opposite: *Narcissus* cv. Divison 1 'Trumpet' – Daffodil (Amaryllidaceae) –
vertical section through the flower.

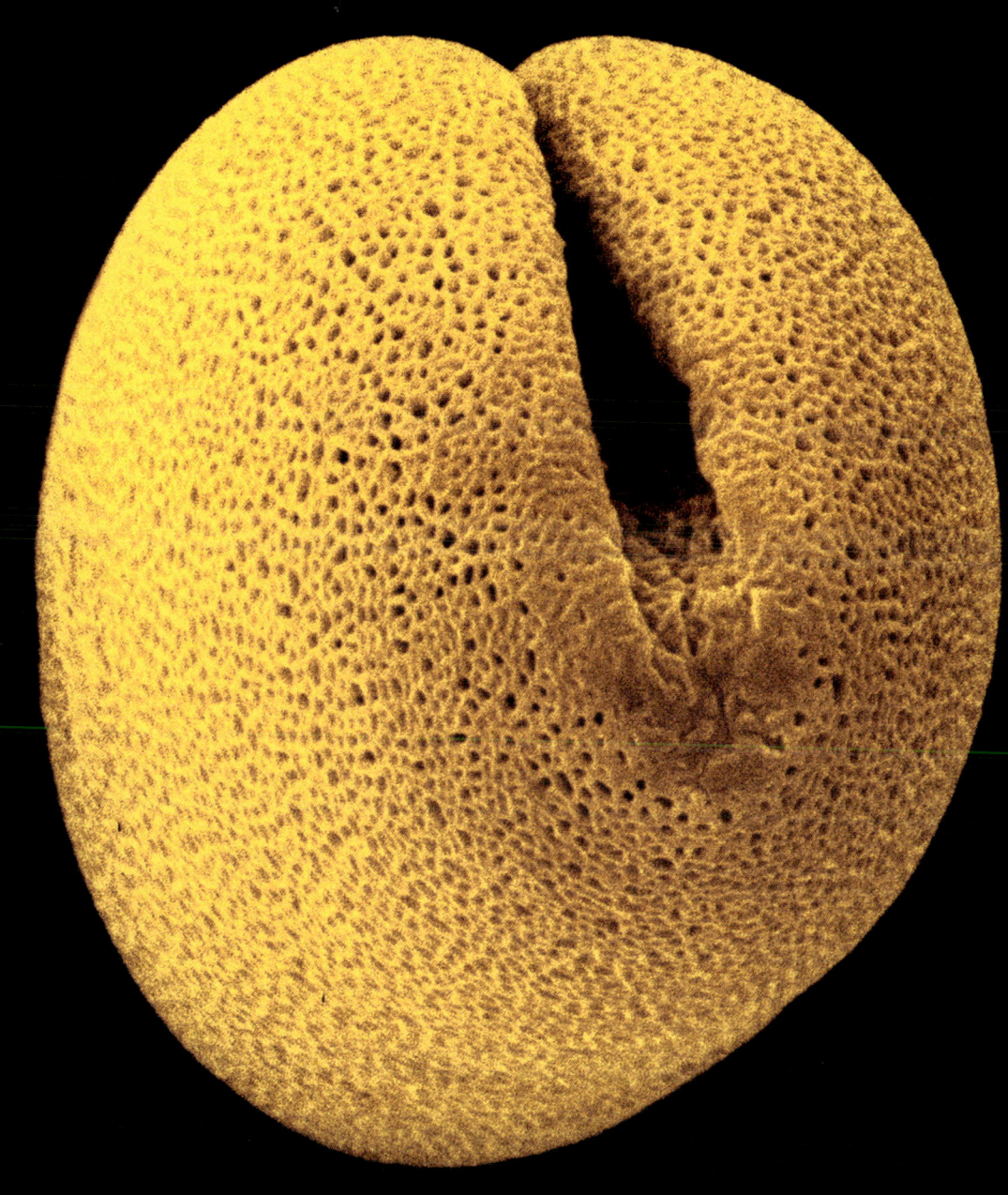

Narcissus cv. – Daffodil (Amaryllidaceae) –
dehydrated pollen grain seen from one end, showing the infolded
aperture. [SEM x 3000]

opposite: *Narcissus* cv. – Daffodil (Amaryllidaceae) –
view inside the trumpet-like flower.

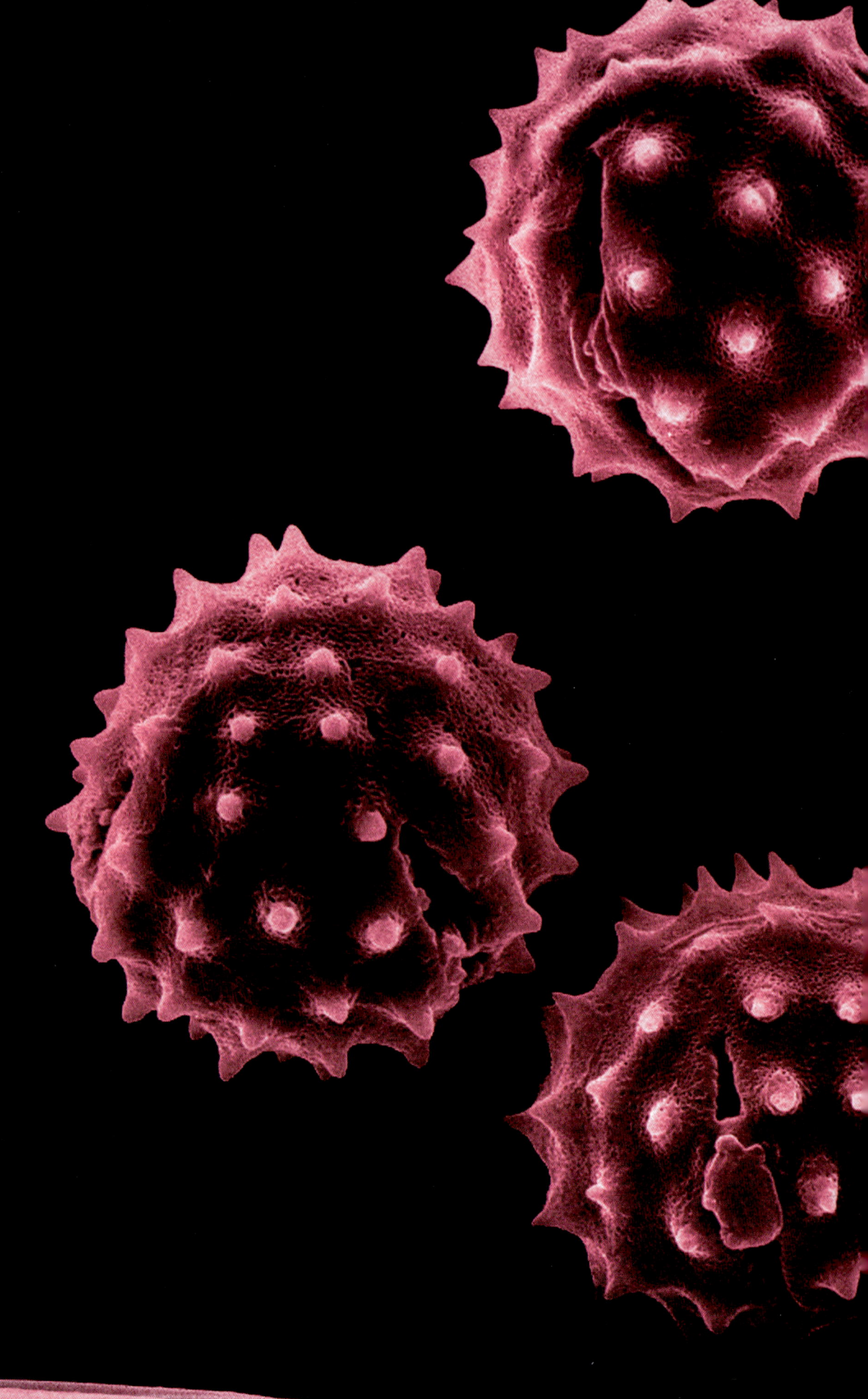

On Matisse

The past tense that he uses to express the future
He will never tire of wondering at wonders
He sleeps so little
And they wake him
At all hours
And from far away there comes
A Wise King called the Wind
To lay at his feet the pollen of the future.

Henri Matisse in his Hundredth Year
Aragon, 1968

Cirsium rivulare ssp. *atropurpureum* (Compositae) –
pollen grains [SEM x 600]

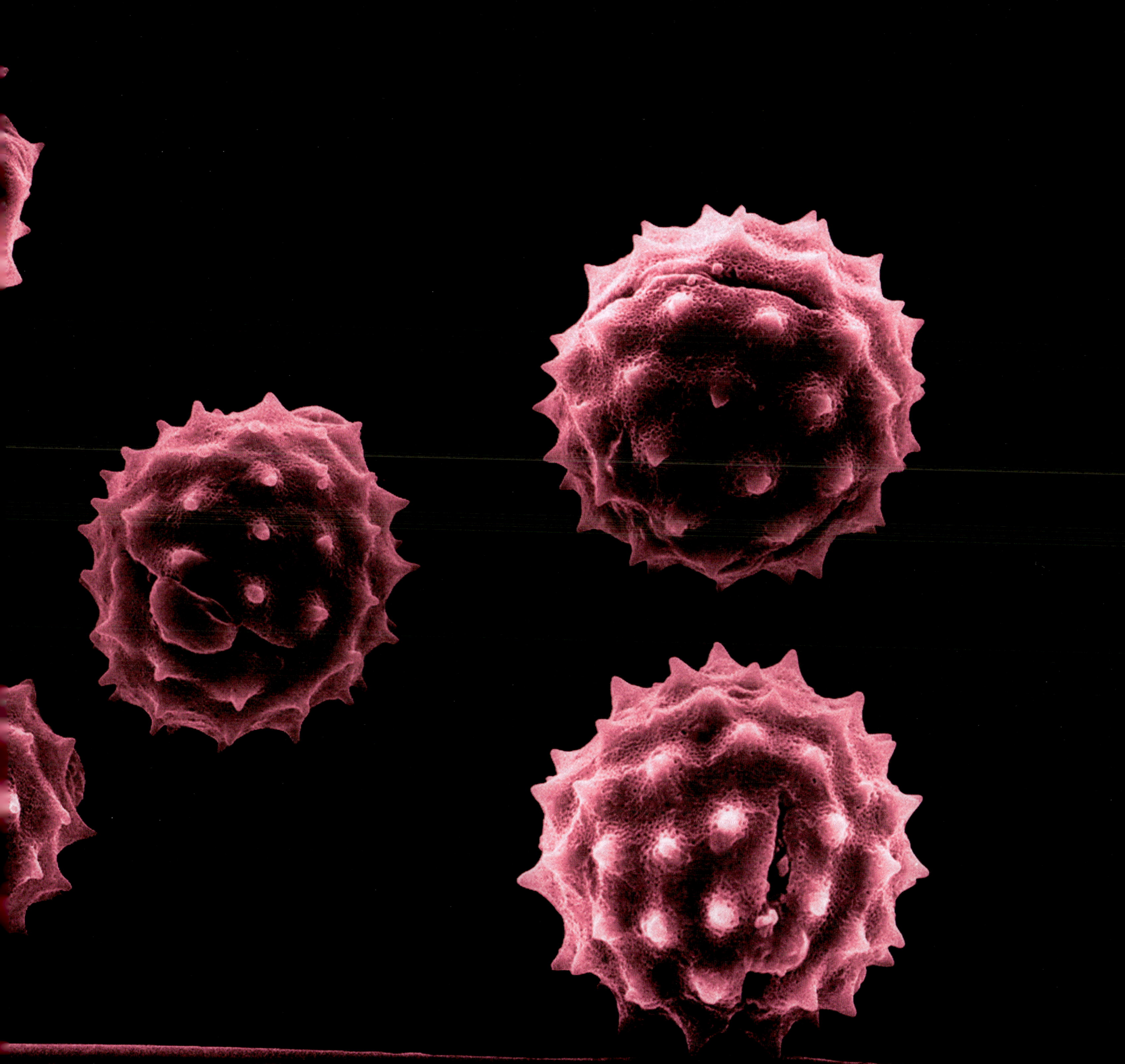

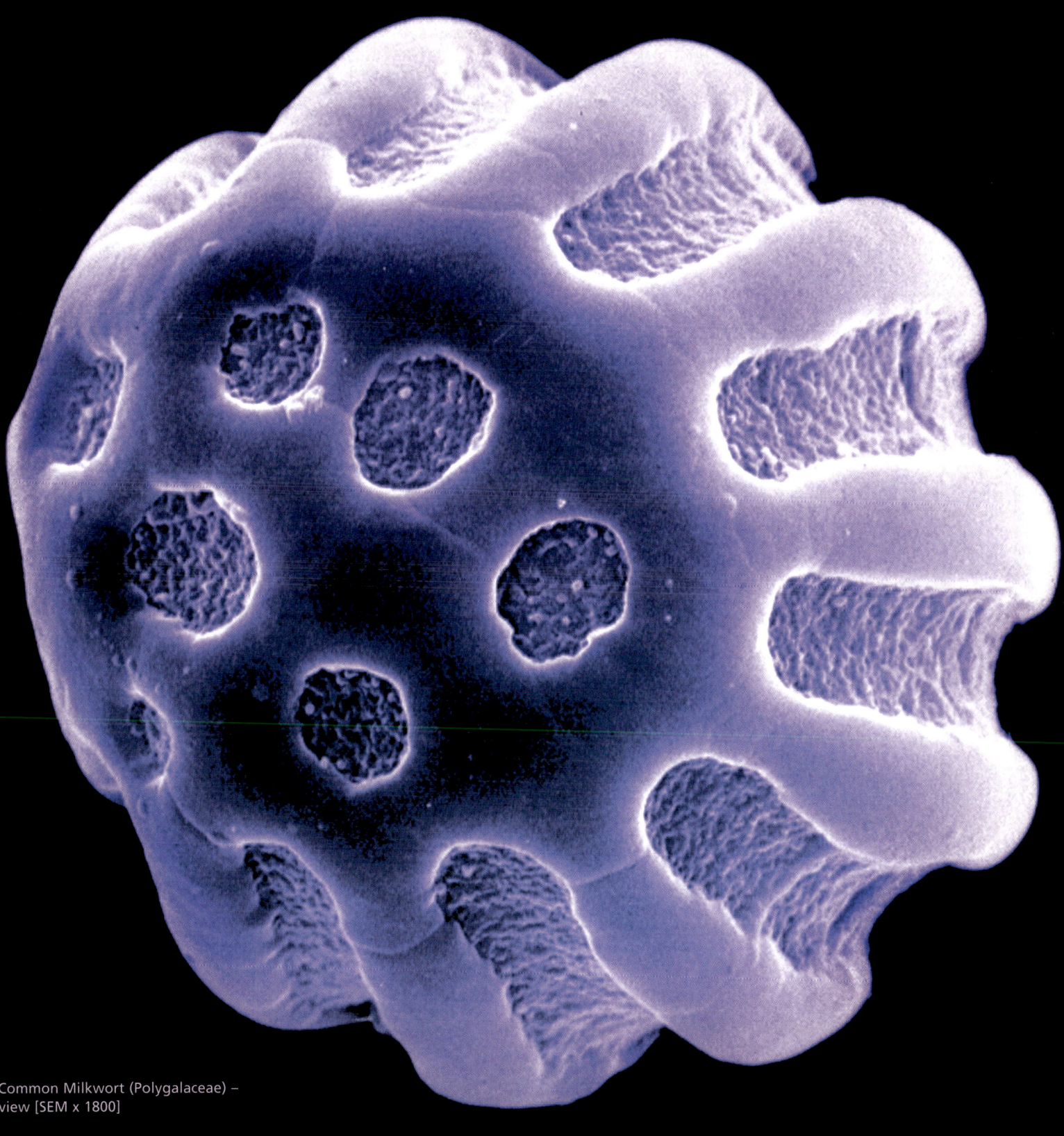

Polygala vulgaris – Common Milkwort (Polygalaceae) –
pollen grain, polar view [SEM x 1800]

opposite: *Polygala vulgaris* – Common Milkwort (Polygalaceae) –
part of an inflorescence

237

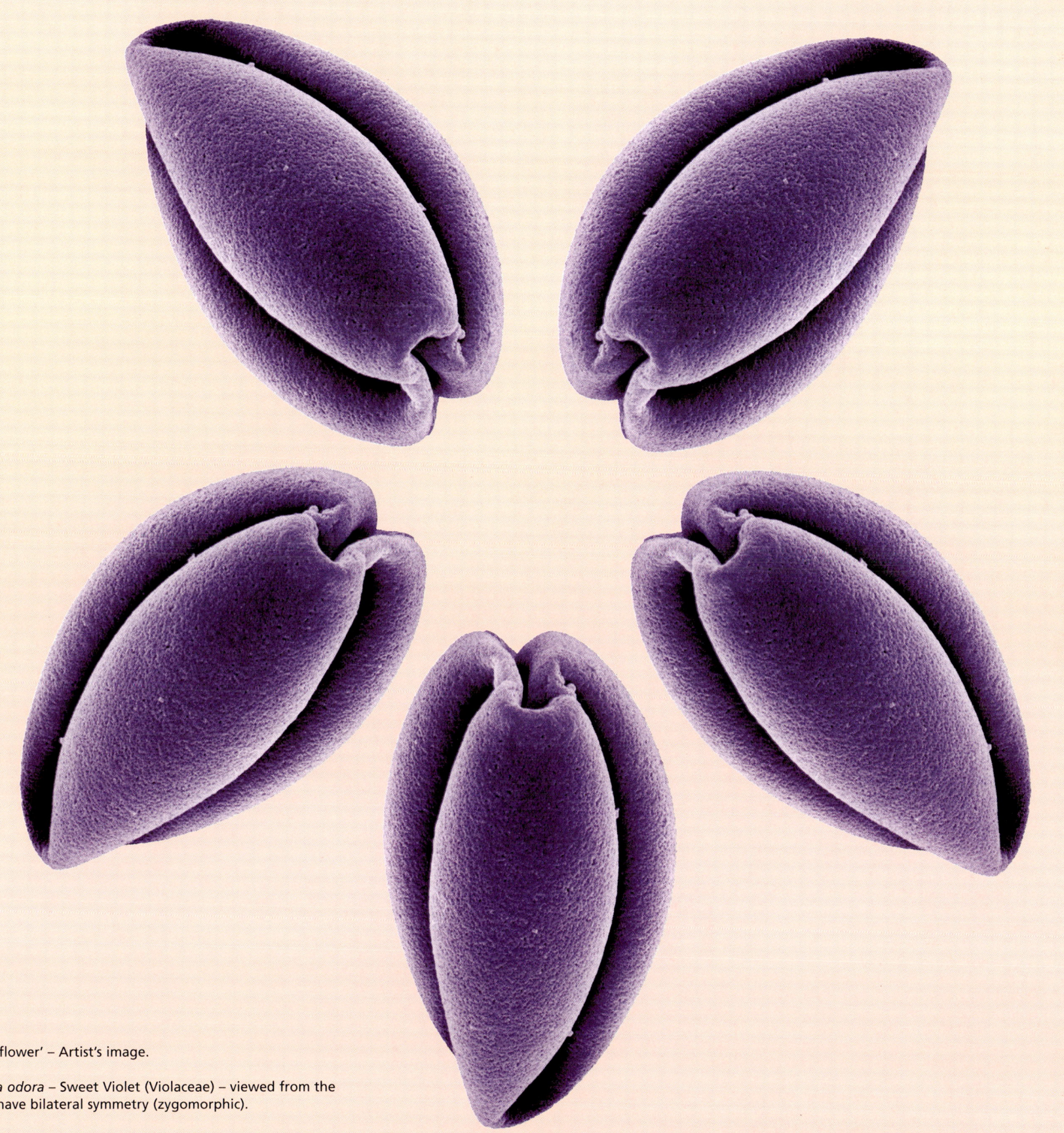

Violet 'pollen flower' – Artist's image.

opposite: *Viola odora* – Sweet Violet (Violaceae) – viewed from the front. Violets have bilateral symmetry (zygomorphic).

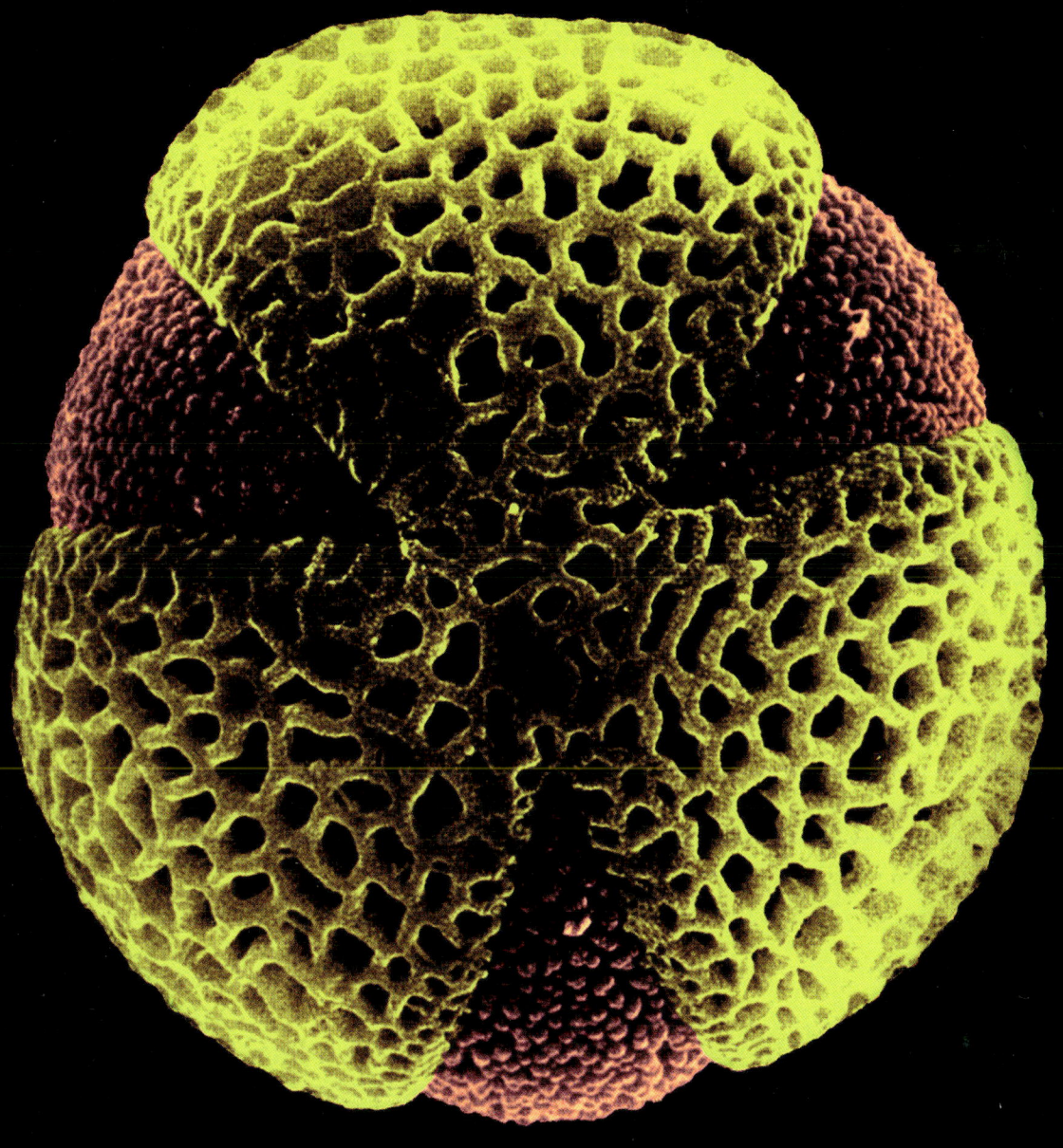

Helleborus orientalis – Hellebore (Ranunculaceae) –
pollen grain in polar view. [SEM x 2000]

opposite: *Helleborus orientalis* – Hellebore (Ranunculaceae) –
close up of flower; note the unusual arrangement of very large showy
sepals surrounding the tiny incurved, nectar-rich petals adjacent to
the stamens. The anthers do not all open at the same time.

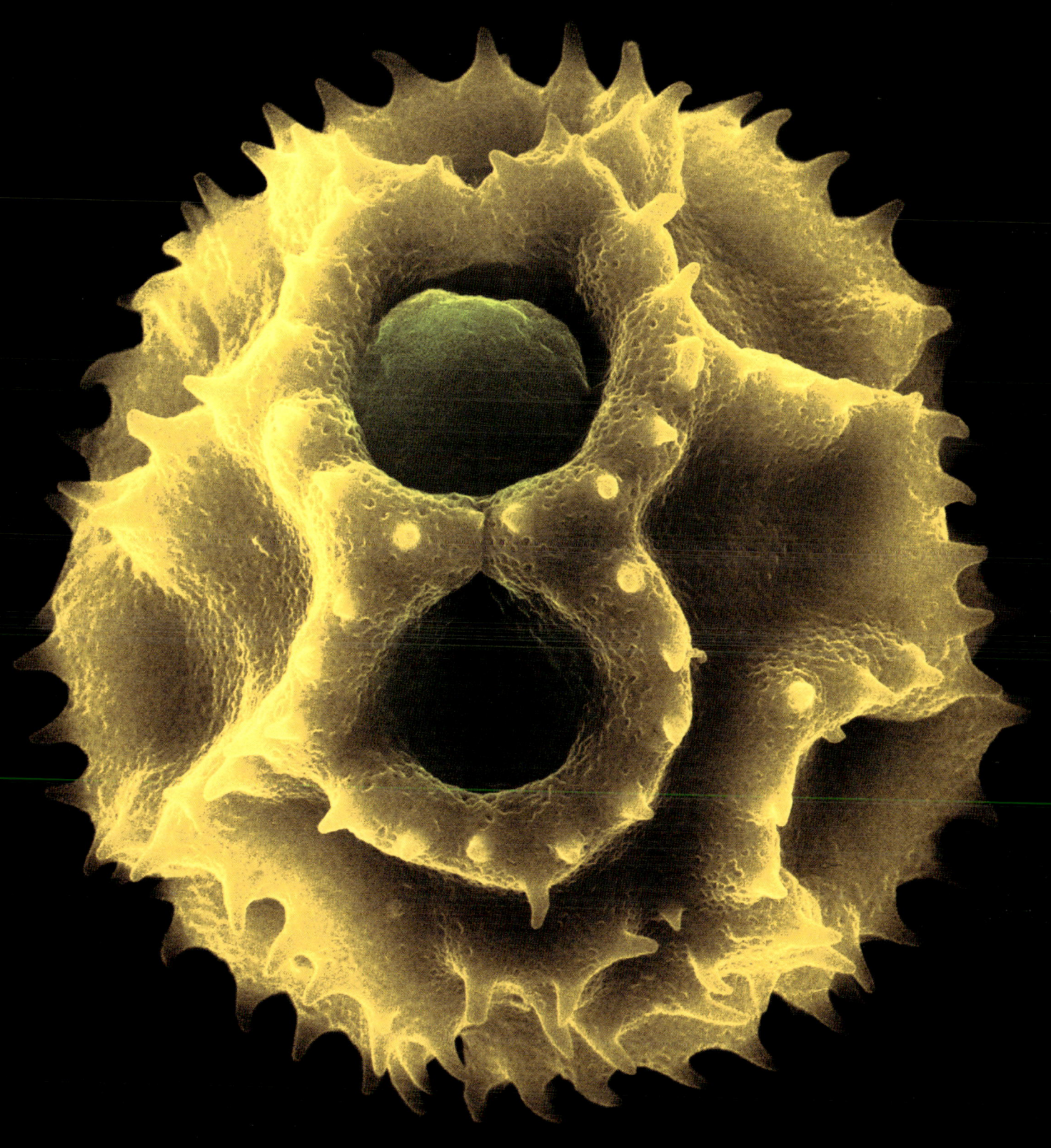

Hieraceum pilosella – Mouse-ear Hawkweed (Compositae) – pollen grain in equatorial view. [SEM x 2000]

opposite: *Hieraceum pilosella* – Mouse-ear Hawkweed (Compositae) – composite flower from above.

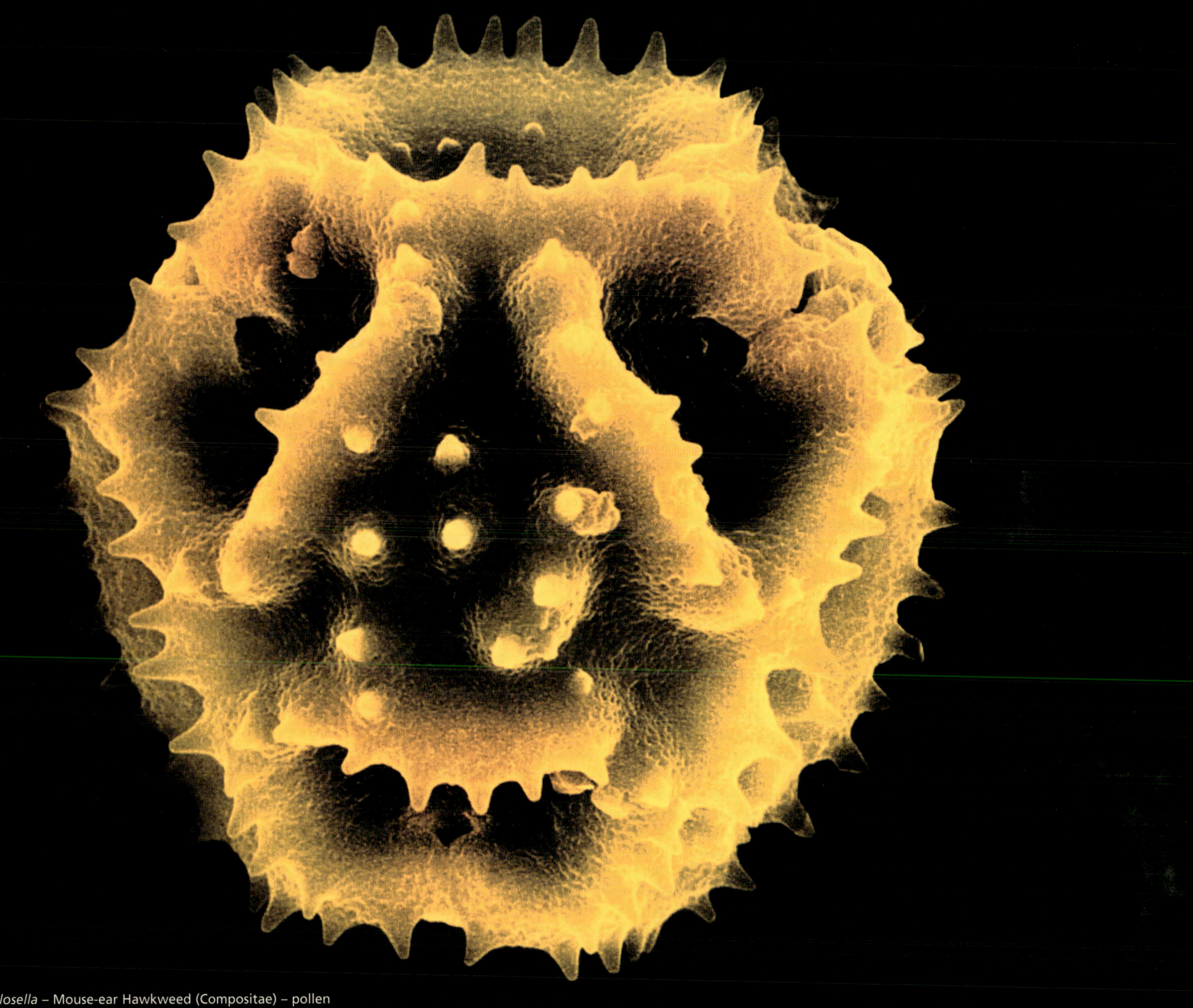

Hieraceum pilosella – Mouse-ear Hawkweed (Compositae) – pollen grain in polar view [SEM x 2000]

opposite: *Hieraceum pilosella* – Mouse-ear Hawkweed (Compositae) – composite flower viewed from the side.

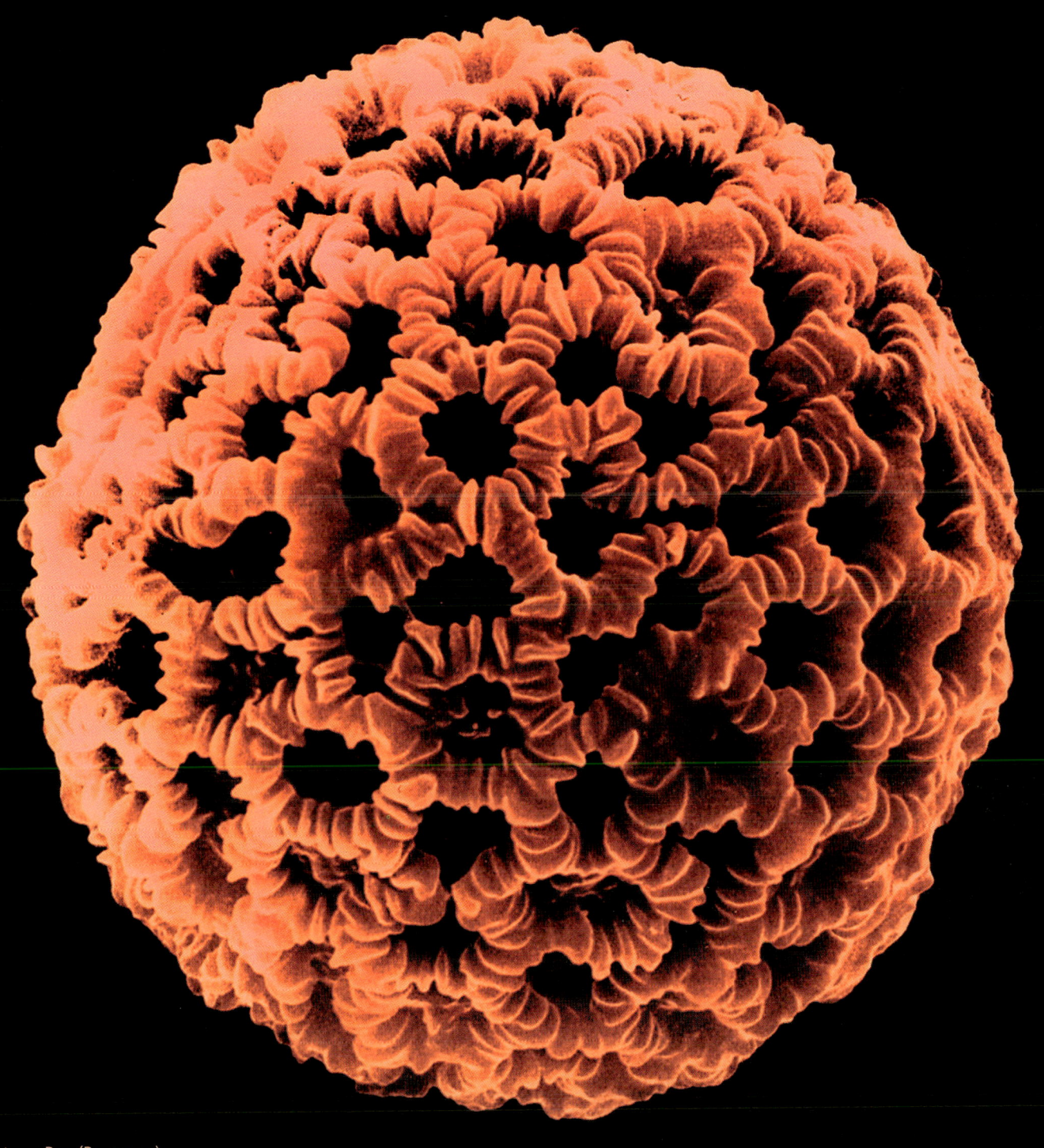

Sarcococca confusa – Christmas Box (Buxaceae) –
pollen grain [SEM x 2000]

opposite: *Sarcococca confusa* – Christmas Box (Buxaceae) –
close up of flower showing clustered stamens.

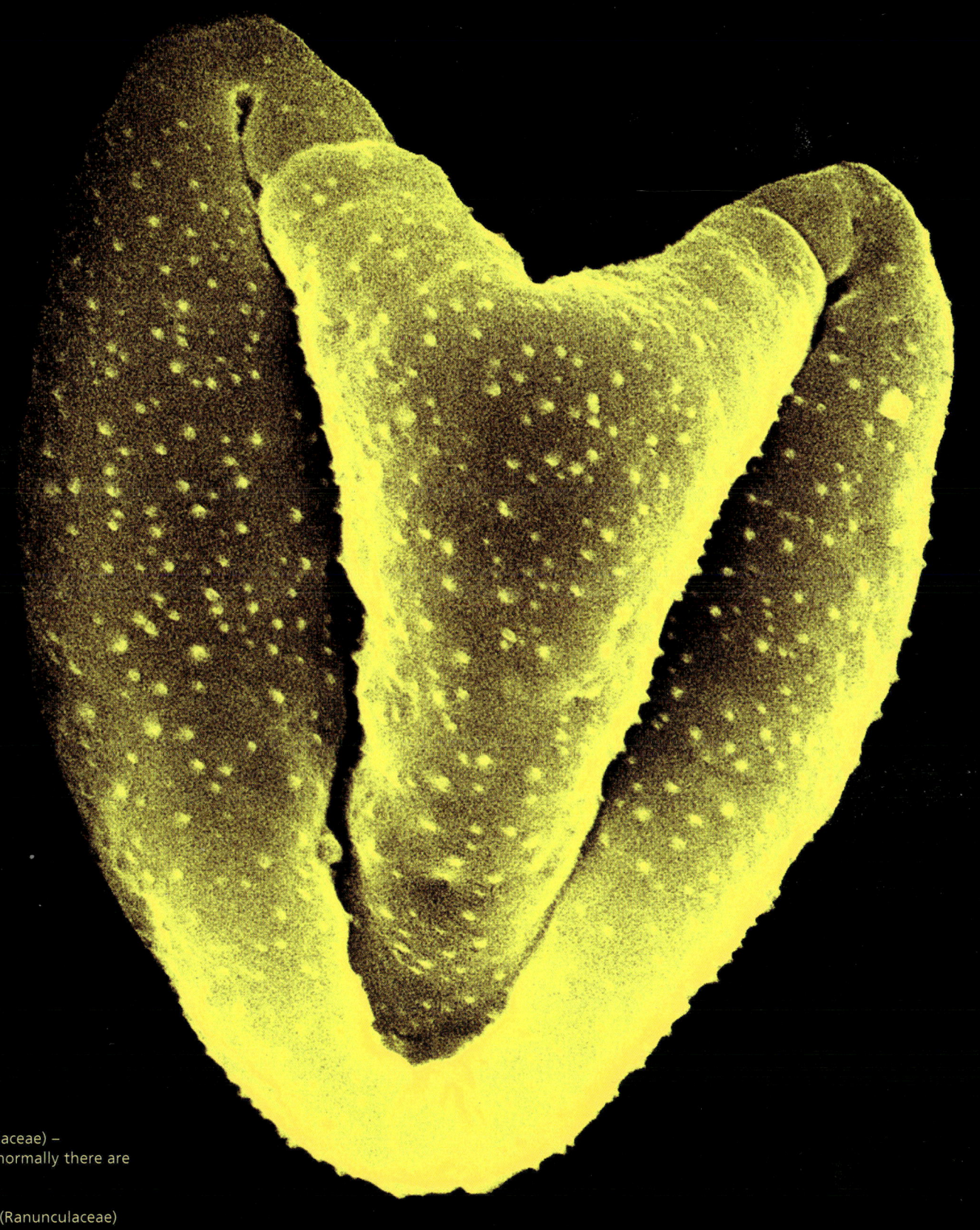

Ranunculus ficaria – Lesser Celandine (Ranunculaceae) –
a pollen grain that has developed abnormally; normally there are
three distinct apertures.

opposite: *Ranunculus ficaria* – Lesser Celandine (Ranunculaceae)

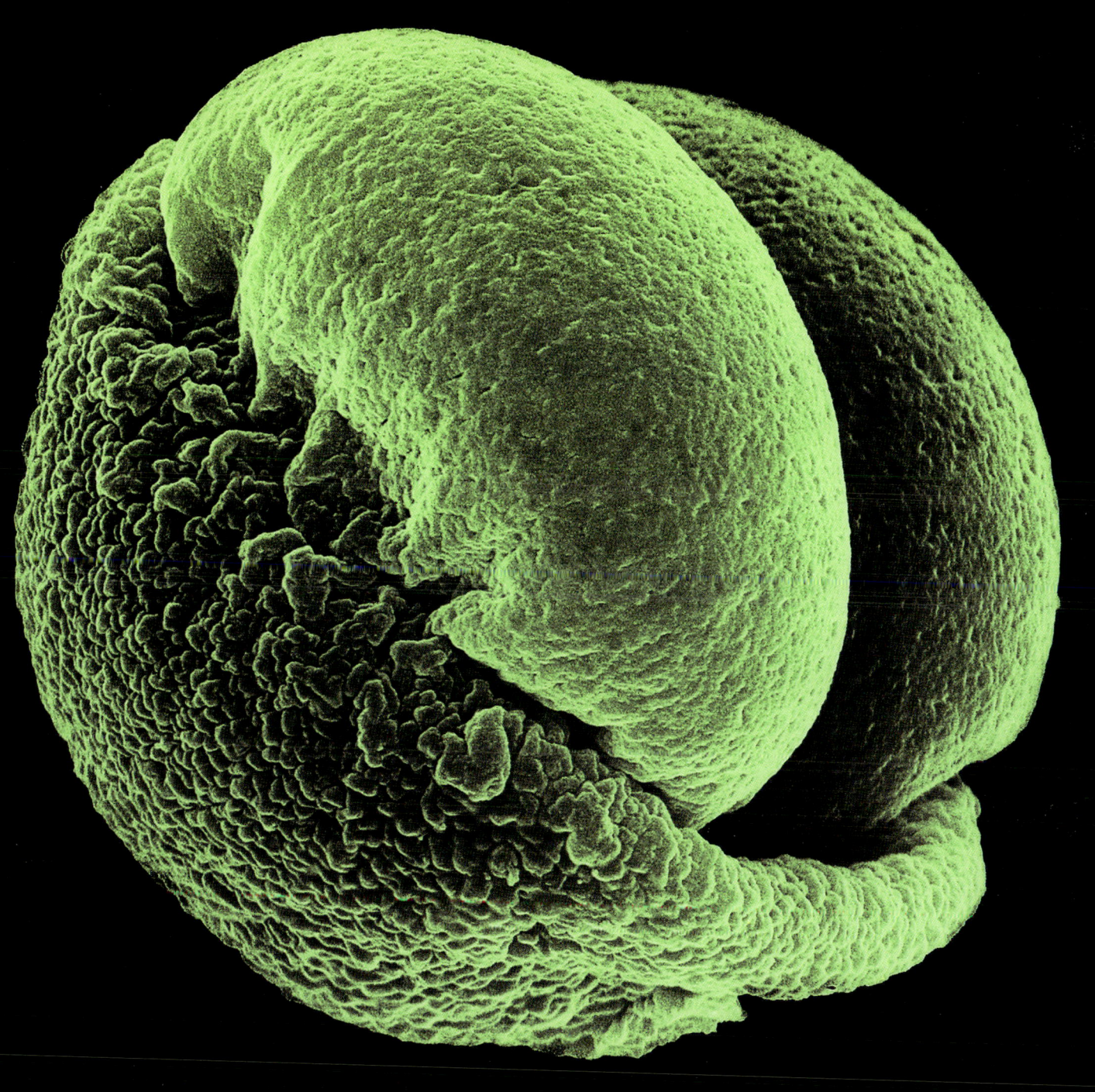

Pseudotsuga menziesii – Douglas fir (Pinaceae) – pollen grain, in unexpanded condition; note the air sacs. [SEM x 2000]

opposite: *Pinus tabuliformis* – Chinese red pine (Pinaceae)

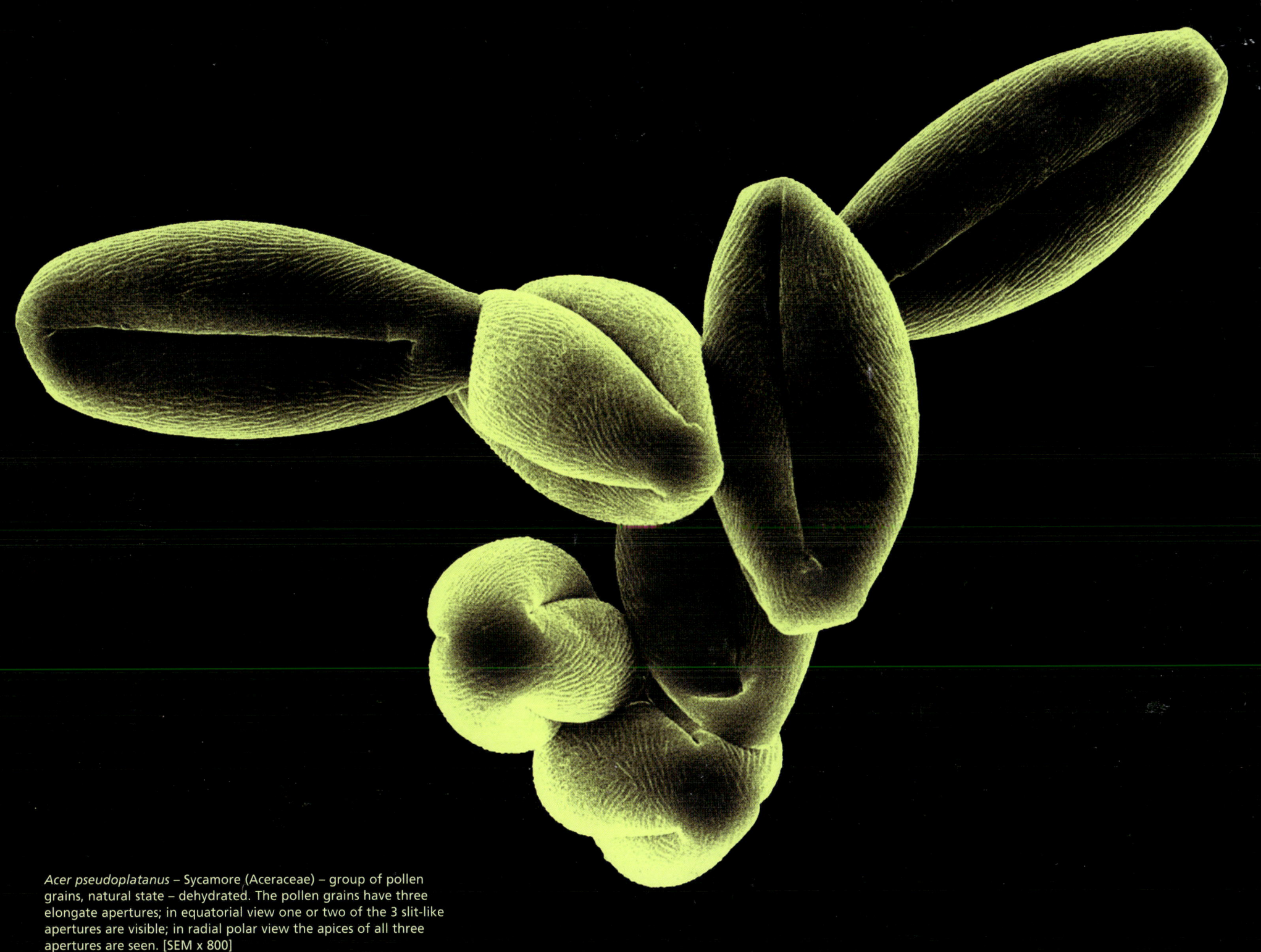

Acer pseudoplatanus – Sycamore (Aceraceae) – group of pollen grains, natural state – dehydrated. The pollen grains have three elongate apertures; in equatorial view one or two of the 3 slit-like apertures are visible; in radial polar view the apices of all three apertures are seen. [SEM x 800]

opposite: *Acer pseudoplatanus* – Sycamore (Aceraceae) – winged seeds developing

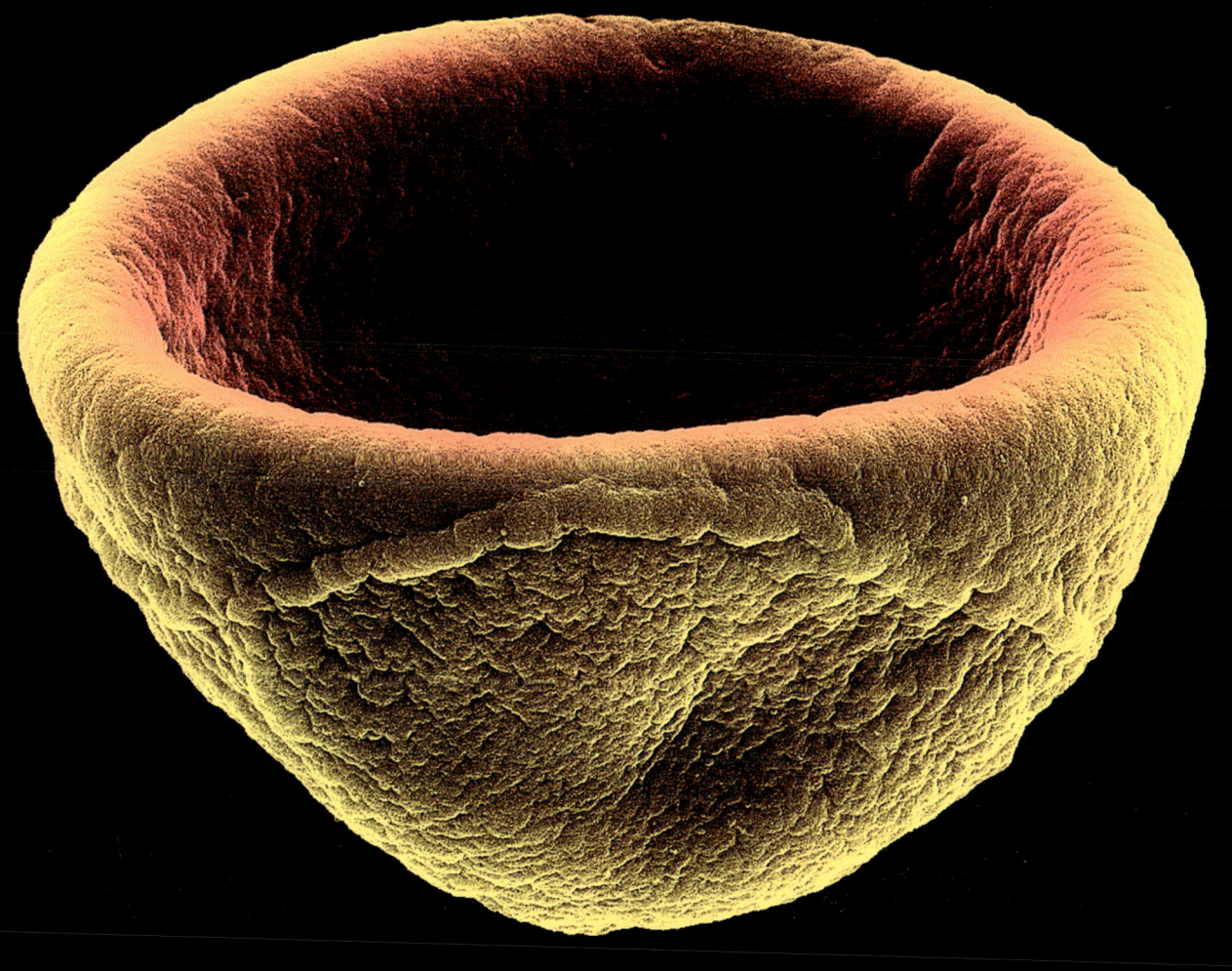

Larix decidua – Larch (Pinaceae) – pollen grain, dehydrated and infolded, forming a bowl. [SEM x 1500]

opposite: *Larix decidua* – Larch (Pinaceae) – male cones (bottom and centre left) and female cone (centre right)

POSTSCRIPT

ROB KESSELER

To the outsider collaborations between artists and scientists may suggest outcomes that result in a hybrid fusion of their respective cultures, raising unrealistic expectations of a super progeny. However, in reality the outcomes are, like pollen, more subtle, far more diverse, and more widely dispersed than might be imagined. Additionally people are prone to assume that all the benefits of such a union are loaded on the side of the artist, and may ask what is in it for the scientist. The artist Joseph Beuys believed in a locus of artistic sensibility centred within every individual and for the possibility of a harmonious union with nature through art. The scientist does well to remind us of the locus of scientific sensibility we also possess.

The production of this book has been a true collaboration, an equal partnership at the interstices between two cultures, the point at which our respective practices overlap. What you see is a distillation of several years' work collecting, preparing and studying plant material; it is also the product of many hours spent in conversation provoked by the differing attitudes we each have to selecting and describing plant material. Differences born out of the practices and conventions of science or of art are differences that need to be acknowledged but also, sometimes, these differences should be challenged.

The paired of images of pollen and flowers in the previous chapter perhaps best exemplify the differences in our approach. The scientist will select the most perfect specimen to give the fullest information with absolute clarity for any chosen sample. It could be argued that artists do the same except that their choices are driven by very different intentions. I often take pollen samples from fresh flowers which, after drying, are transferred directly to the microscope where the pollen grains are revealed in a variety of states from fully hydrated to collapsed, imploded or fractured – 'natural' sculptures on an infinitesimally small scale. In each case I have carefully selected pollen grains to reflect the flowers from which they came. The colour, heightened at times to a bizarrely unnatural state, provides a chromatic interference to draw attention to the form or functional aspects of the pollen grain. It may refer to the natural colour of the pollen, or the colour of the flower from which it was taken, or it might be my wilfully unscientific meddling with nature – ornamental fantasy. I have attempted to create images that lie somewhere between science and symbolism, in which the many complexities of artistic representation of plants are concentrated into mesmeric visual statements, analogous morphologies with the power to burn themselves into the memory

Flying Pollen: a series of banners created for 'Go Wild' – a festival of bio-diversity held at Kew Gardens during the summer of 2003. Each banner, featuring a different pollen grain hugely magnified, was stretched between trees in wooded areas of the gardens, close to the locations from where the pollen was originally collected from native wild flowers. The coarse mesh of the banners became analogous to a fine filter for trapping airborne particles – in this case pollen grains.

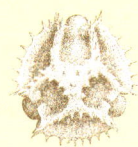

APPENDICES

Apis mellifera – Honey Bee (order Hymenoptera) – visiting a flower of *Taraxacum officinale* – Dandelion (Compositae). Note the pistils with branched stigmas which emerge from a ring of united stamens; the stigma branches, which are covered in pollen, curl back as they mature. Note also the amount of pollen on the body and legs of the bee.

GLOSSARY

Acetolysis – a process developed during the 1930s by the 'Father' of modern comparative pollen morphology, Gunnar Erdtman. Acetolysis was developed to clear the pollen of the internal cellular fraction, as well as the intine, and any external lipidic coatings (Erdtman, 1936). In light microscopy, the morphological details of the pollen exine are then clearly visible, and comparable to fossilised pollen, which also retains only the pollen exine. The acetolysis mixture comprises nine parts acetic anhydride and one part sulphuric acid. The pollen grains are immersed in the mixture, in tubes, and heated at 100° in a hot water bath, or dry heater for anything from 1-10 minutes, depending on the fragility or toughness of the exine involved. (This process should only be undertaken within a fume hood in laboratory conditions.)

Allergenic reaction – a susceptible individual is exposed to an allergen. In the initial reaction the body responds by producing immunoglobulin E (Ig E) in antibody forming cells of lymphoid tissues. Immunoglobulin E circulates in the serum within the bloodstream where it has an affinity for certain epithelial and mucosal cells (basophil and mast cells). The immunoglobulin E antibody molecules attach to these cells by a footpiece (the Fc region of the molecule) where they may remain bound for periods of several weeks. The immunoglobulin E antibody molecules are vastly smaller than the mast cells; each mast cell may have as many as 100,000 Ig E molecules on its surface. Each Ig E molecule has two arms with a terminal recognition site for its specific allergen. It is in communication with the mast cell membrane through the membrane glycoprotein to which it is attached. The next time the susceptible individual encounters the allergen it binds to pairs of adjacent Ig E molecules on the mast cell surface. The binding inter-action triggers the rapid release of tissue mediators (pharmacologically active substances from granules secreted by the mast cell including histamines and enzymes) which bring about the symptoms of allergic reaction.

Androecium – collective term for the stamens within a flower, the male reproductive organs (> 'gynoecium', 'stamens').

Angiosperm – flowering plants almost all of which produce seeds enclosed within carpels. They range from weedy annuals, perennials, bulbous plants, shrubs to large trees. They are unique in the plant kingdom in undergoing the double fertilisation event. The angiosperms comprise two major groups of flowering plants: 'monocotyledons' and 'dicotyledons' (>'gymnosperms').

Anther – apical portion of stamen, usually comprising four pollen sacs (locules), which produces the pollen grains (> 'filament', 'locules').

Anthocyanin – any of a group of glycoside pigments formed by the addition of sugars and other residues to an anthocyanidin pre-cursor, usually pelargonidin, delphinidin or cyanidin.

Aphids – minute insects in the order Hemiptera ('True Bugs') possessing piercing mouth parts adapted for sucking the juices of plants and other animals. The body is typically pear-shaped with narrow head and bulbous abdomen. Greens and browns the predominant colours. Wings when present are usually clear and membranous. However, polymorphism is common and wingless forms occur in most species. Some species of aphid, the scale bugs (superfamily Coccoidea) with often wingless and legless females, exude a sticky 'honeydew' which solidifies into sugary lumps as the water evaporates (> 'honeydew').

Bee bread – a compressed mixture of nectar and pollen used as brood food by bees.

Beeswax – the wax for building the combs. It is secreted in small flakes by the worker bees from special glands located on the underside of the body, between the abdominal segments (> 'combs').

Bract – a leaf-like organ subtending an inflorescence (> 'spathe').

Calyx – the sepals collectively.

Carotenoid – yellow, orange, or red fat soluble pigments found in all photosynthesising cells, where they act as accessory pigments in photo-synthesis. They are also found in other organs, for example roots, petals and pollen grains.

Carpel – the structure that holds and encloses the ovules in flowering plants; a single carpel is synonymous with a 'pistil' or an 'apocarpous gynoecium'. A group of carpels is termed a 'syncarpous gynoecium'.

Chromatin – the complex of proteins, DNA and small amounts of RNA, of which chromosomes are composed (> 'chromosomes').

Chromosomes – thread-like structures in the eukaryotic cells composed of chromatin; they carry genetic information. During cell division they condense and become visible under the microscope (> 'chromatin', 'eukaryote', 'nucleus').

Chalaza/chalazal area – the basal region of the ovule where the nucellus and integuments fuse. It may or may not coincide with the position of the funiculus depending on the mode of ovule orientation (> 'funicle').

Combs – the layered complex of hexagonal wax cells built by worker bees to contain bee larvae,or nectar (honey) or pollen.

Corbicula/corbiculae – the pollen basket – stiff hairs (combs) on the hind legs of bees, where pollen is collected and pressed into pollen loads (> 'pollen loads').

Corolla – the petals collectively.

Corolla tube – a tube is formed when the margins of the individual petals are completely or partially fused. In some species the tube may be very long and/or very narrow.

Cotyledon – the first leaf, or leaves, of the embryo in seed plants.

Cretaceous – a geological period from about 140 to 65 million years before the present. It is divided into two main epochs – the Lower or Early and the Upper or Late. The flowering plants evolved in the Early Cretaceous, and the Dinosaurs became extinct at the end of the Late Cretaceous.

Deciduous – woody or perennial plants that shed their leaves before winter, or dry season, an adaptation that reduces water loss.

Dicotyledons/dicotyledonous – one of the two major groups of flowering plants, the other being monocotyledons. The dicotyledons are named because the germinating embryos (seeds) usually have two cotyledons. They may be either herbaceous, shrubby or trees. Other distinguishing characters include flower parts, usually in fours or fives, or multiples of, vascular bundles in a ring, and a primary root that develops into a tap root (> monocotyledons').

Dioecious - with separate male and female plants. From the Greek, meaning living in separate houses (> 'monoecious').

Diploid – a nucleus or individual having twice the haploid number of chromosomes in the nuclei of its somatic cells (> 'haploid', 'gamete/ gametophyte', 'meiosis', 'somatic').

Egg cell – the haploid female cell which will be fertilised by a male generative cell to form a diploid zygote (> 'generative cell', 'zygote').

Electron microscope – a microscope where a parallel beam of electrons, from an electron gun – rather than light – is used to illuminate the subject via a series of magnetic lenses. There are two types of electron microscope: a scanning electron microscope – where electrons are bounced back off a thick section, or complete subject, and collected by an electron detector, of which there are various types, the most commonly used in biology being a secondary electron detector. The image produced is three-dimensional. In the other type, a transmission electron microscope, the electrons penetrate through an ultrathin film (nanometres thick) of the subject. The resolving power of modern electron microscopes is in excess of 50,000 times, higher for non-biological material (> light microscope).

Embryo sac – large oval cell in nucellus of ovule of flowering plants in which fertilisation of egg and development of embryo occur.

Endosperm – storage tissue in the seeds of most flowering plants, but not in any other seed plants. It is a compact triploid tissue without intercellular spaces; starch, hemicelluloses, proteins, oils and fats are stored in endosperm.

Equatorial view – this is a confusing topic for the beginner, because it relates to the type of tetrad from which the mature pollen has developed. Typically, in pollen grains with one elongate aperture, it is when the pollen grain, viewed from its long or short axis, has the aperture only half visible, while in pollen grains with three elongate apertures such as in Hellebores or Witch Hazels, it is when only one or two of the three apertures are completely visible (page 57). (> 'polarity', 'polar view', 'tetrad'; see also Erdtman 1943, 1952 or 1969).

Eukaryote/eukaryotic – an organism whose cells have a nucleus (> 'chromosomes', 'nucleus').

Exospore – the outer, usually acid/oxidation resistant, layer of a spore wall, the inner layer being the 'endospore'. Another layer, the perispore (synonym 'perine'), is present in some fern groups. It lies outside the exospore, but is not normally resistant to acids and oxidation.

Families of plants – plants, like other biological organisms, are grouped into 'taxonomic hierarchies'. For flowering plants the main groups are: angiosperms; orders; families; genera; species (> 'taxonomy').

Filament – (in relation to stamen and anther) the stalk of a stamen bearing the anther at its apex (> 'stamen', 'anther').

Flavones – a group of flavonoid pigments the members of which contain an unaltered flavonoid nucleus. (> 'anthocyanin', 'flavonoids').

Flavonoids – a group of plant compounds all of which contain a two-phenylbenzopyran nucleus. They include the flavones, anthocyanins, flavanones, chalcones, aurones and flavonols.

Funiculus/funicle – the stalk attaching the ovule, and later the seed, to the placenta or ovary wall in angiosperms. It serves as an anchor and provides a vascular supply to the ovary and seed.

Galls – an abnormal localised swelling or outgrowth produced by a plant as a result of attack by a parasite. They may be caused by bacteria, fungi, nematodes, insects or mites.

Gamete/gametophyte – a cell or nucleus that may participate in sexual fusion to form a zygote. It is normally haploid and thus on fusion of two gametes a diploid zygote is formed.

Generative cell – a haploid male cell which divides to form two male gametes one of which will fuse with the female egg cell to form a diploid zygote, and the other with the polar nuclei in the embryo sac to form the triploid nutritive endosperm tissue (> 'egg cell', 'endosperm', 'embryo sac', 'haploid', 'diploid', 'zygote').

Glycoside – a compound formed by the reaction of a pyranose sugar with a non-sugar molecule (an aliphatic or aromatic hydrocarbon) termed the aglycone.

Gondwana – the southerly of the two ancient continents into which the earth's land mass, Pangaea, was once divided (> 'Laurasia').

Gymnosperms – seed plants that differ from angiosperms in having naked seeds with no carpellary structure and no double fertilisation event. Gymnosperms include Araucarias, Conifers, Cycads, Ephedras, Gingko, Gnetaceae and *Welwitschia* (> 'angiosperms').

Gynoecium – the female organs of a flower, comprising one or more pistils (stigma + style + ovary). Where there is only one pistil, 'carpel', 'gynoecium' or 'syncarpous gynoecium' are synonymous.

Haploid – a nucleus or individual containing only one representative of each chromosome of the chromosome complement (> 'diploid').

Haploid/haplophytic generation – the gametophyte.

Herbarium – a collection of pressed and dried plants, labelled and systematically stored in a damp and pest free environment, for scientific study. The Herbarium of the Royal Botanic Gardens, Kew, with at least eight million specimens, is one of the largest in the world.

Hermaphrodite – bi-sexual; of an individual plant or animal having both male and female gametes.

Honeydew – a sticky, sugary exudate produced by certain aphids, in particular scale bugs, which solidifies into sugary lumps as the water evaporates (> 'aphids').

Incompatibility – in flowering plants, failure to fertilise and subsequently set seed after pollination has occurred (> 'pollination').

Inflorescence – a branching arrangement of flowers on the floral axis. The flowers may be highly condensed (e.g. daisy), or loosely grouped (e.g. lilac).

Integument – a protective structure (envelope) that develops from the base of an ovule, enclosing it almost entirely with the exception of the micropyle (> 'micropyle', 'nucellus').

IQ (intelligence quotient) – A number arrived at by means of intelligence tests, and intended to express the degree of intelligence of an individual in relation to the average for the age-group, which is fixed at 100. Introduced by the German psychologist, William Stern, in 1912, and widely adopted during much of the twentieth century, it is now considered outdated in some of its concepts and assumptions.

Isoflavones - isomers of flavones in which the B group of the flavonoid nucleus is attached to the third rather than the second carbon of the central C3 group. They are particularly common in the pea family (Leguminosae).

Labellum – the usually enlarged adaxial (lower) petal of an orchid flower.

Laurasia - the northerly of the two ancient continents into which the earth's land mass, Pangaea, was once divided (> 'Gondwana').

Light microscope – a microscope where the light is focussed on the subject via a series of precision ground glass lenses; the resolution is between 100 and 1500x (> 'electron microscopy').

Locule – a cavity within which specialised organs may develop. For example, the anther locule or the ovary locule (> 'anther', 'ovary').

Mandibular gland – a pheromone secreting gland lying immediately above the jaws of the bee (>'pheromone', 'queen bee substance').

Meiosis – reduction division of the diploid cell to form four haploid cells (> 'mitosis', 'zygote').

Micron/micrometre – a thousandth of a millimetre (abbreviation μ) (> 'nanometre').

Micropyle – the small channel that remains between the tips of the integuments at the apex of the ovule. Usually the pollen tube enters the nucellus via the micropyle. Sometimes the micropyle persists as a small pore in the seed, through which water is absorbed before and during germination (> 'integument', 'nucellus', 'ovule', 'pollen tube').

Mitosis – the process by which a cell divides to form two 'daughter' cells, each having a nucleus containing the same number of chromosomes, with the same genetic composition as the original cell (> 'meiosis').

Monocotyledons/monocotyledonous – one of the two major groups of flowering plants, the other being dicotyledons. The monocotyledons are named because the germinating embryos (seeds) usually have one cotyledon. Most are herbaceous, the stems lack secondary thickening and true trees do not occur, although some groups such as the palms have arborescent forms. Other distinguishing characters include flower parts usually in threes, or multiples of, scattered vascular bundles, and a fibrous root system (> 'dicotyledons').

Monoecious – with separate male and female flowers on the same plant. From the Greek, meaning living in the same house (> 'dioecious').

Morphology – the study of form, particularly of external structures – the introduction of the word is attributed to the poet Johann Wolfgang von Goethe (1749-1832), who has also made very important con-tributions to botany and other branches of the natural sciences (see Bibliography).

Nanometre – one thousandth of a micron/micrometer (> 'micron/micrometre').

Nectar – a sugary solution secreted by nectaries, most commonly in the flowers of insect or bird-pollinated plants. It acts as an attractant and reward (> 'nectaries').

Nectaries – glands secreting nectar. Usually situated in the base of the flower or spur (e.g. Aquilegia) to attract pollinators (> 'nectar').

Nucellus – the central tissue of the ovule containing the embryo sac and surrounded by the integuments (> 'embryo sac', 'integument', 'ovule').

Nucleus – the part of the eukaryotic cell that contains the genetic material (> 'chromosomes', 'eukaryote').

Organelles – a membrane enclosed structure in the cytoplasm organised to carry out a specific process.

Ovary – the swollen basal region of a pistil (cf. 'carpel') containing one or more ovules (> 'ovule', 'pistil', 'carpel').

Ovule – the female gamete structure of seed plants (angiosperms, gymnosperms), which develops into a seed after fertilisation of its egg cell (> 'seed plant').

Parthenogenesis – 'virgin birth', the development of an embryo into an egg cell without fertilisation. It can occur in plants or animals, and can be haploid or diploid. In haploid parthenogenesis (e.g. in the drones of Apis mellifera) the eggs are produced by meiosis in the usual way, and are therefore, haploid. These develop into a new individual whose cells are, therefore, haploid – effectively a cloned individual. In diploid parthenogenesis the eggs, instead of being formed by meiosis, are formed by mitosis, with the result that they are diploid rather than haploid, the resulting individual will, therefore, have the normal diploid constitution. This happens in certain stages of the life cycle of aphids. It provides a rapid and efficient way of increasing numbers without the necessity of males (>'diploid', 'haploid', 'meiosis', 'mitosis').

Petal – an individual unit of a corolla (> 'corolla').

Pheromone – a chemical substance the release of which by an animal into its surroundings influences the development or behaviour of other individuals of the same species (> 'queen bee substance').

Pistil – an individual carpel comprising an ovary, a style and a stigma.

Placenta – the part of an ovary wall on which the ovules are borne.

Polar nuclei – the pair of nuclei in the ovule with which one of the two male generative nuclei fuse to form the triploid nutritive endo-sperm tissue (> 'egg cell', 'generative cell', 'ovule', 'vegetative cell').

Polarity – refers to the position of the four pollen grains relative to each other during the tetrad stage. There are two poles, the distal pole and the proximal pole. The proximal pole is the midpoint of the face where each of the four young pollen grains is associated with its 'sisters' in the tetrad, while the distal pole is the midpoint of the opposing face of each grain. Pollen apertures usually develop before the pollen grains separate from the tetrad as mature individuals; aperture position and arrangement in each pollen grain follows a genetically pre-destined pattern, which relates to the position and arrangement of the apertures in each of the other three pollen grains during tetrad stage. Viewed with a light or scanning electron microscope the pollen of the majority of flowering plants appears different in different views because of the non-global distribution of the apertures, although there are exceptions such as polyporate pollen grains. The analogy of aperture position, in relation to the globe, to describe the position in which the pollen grain is being viewed relative to the apertures was introduced by Gunnar Erdtman in 1943 (> 'tetrad', 'polar view', 'equatorial view'; see also Erdtman 1943, 1952 or 1969).

Polar view – this is a confusing topic for the beginner, because it relates to the type of tetrad from which the mature pollen has developed. Typically, in pollen grains with one elongate aperture, the polar view is when either the complete aperture is visible, for example lily pollen, or when no part of the aperture can be seen; while in pollen grains where there are three elongate apertures, such as in Hellebores or Witch Hazels, polar view is when all three apertures are partly visible in radial symmetry. (> 'polarity', 'equatorial view', 'tetrad'; see also Erdtman 1943, 1952 or 1969).

Pollenkitt – is comprised mainly of saturated and unsaturated lipids, carotenoids, proteins and carboxylated polysaccharides. It is found in all angiosperms so far studied, but seems to be absent from Bryophytes, Pteridophytes and Gymnosperms. It has various functions: containing the sporophytic proteins inside the exine cavities; keeping pollen grains in or near the anthers until collection by pollination vectors; holding pollen grains in clumps so that they reach the stigma together in larger pollen 'parcels'; allowing adhesion to insect bodies, birds beaks, etc.; protecting the cytoplasm of pollen grains from solar radiation; preventing excessive loss of water from the cytoplasm; determining the colour of pollen; attracting pollinators with oily and perfumed components.

Pollen loads – the pollen collected by bees and compacted into special 'pollen baskets' on their hind legs to be carried back to the hive to make into 'bee bread' (> 'corbicula', 'bee bread').

Pollen tube – the tube which develops from a germinating pollen grain; it carries the male gametes to the embryo sac (> 'embryo sac', 'gamete', 'generative nucleus', 'vegetative nucleus').

Pollinarium (pl. Pollinaria) – the entire male reproductive structure, in most orchids and many species of Asclepiadaceae, which will be removed by a pollinator to be carried to another flower. It comprises the pollinia, sometimes a caudicle (stalk) and a viscidium.

Pollination – transference of pollen grains from one seed plant to the stigma of another plant of the same species. Usually this involves an external agent – animals, wind or water.

Pollinium (pl.Pollinia) – a structure in which the individual pollen grains remain massed together, to be transported as a single unit during pollination. Many members of the families Orchidaceae, and the Asclepiadaceae ('milkweeds') (> 'polyad', 'tetrad').

Polyad – groups of pollen grains which remain attached at maturity and are dispersed as a unit. The groups are usually in multiples of four grains (> 'pollinium', 'tetrad').

Proboscis – a trunk-like feeding apparatus (e.g. in elephant), a sucking organ or elongated mouthpart in some insects.

Propolis – plant resin that bees collect from plant buds and mix with enzymes they produce, as well as beeswax and pollen. They cover all parts of the hive with this for disinfecting purposes. They also use it for sealing cracks and holes or for reducing the size of the hive entrance in winter.

Queen bee substance – pheromone secreted in the mandibular glands of the queen bee (> 'mandibular glands', 'pheromone').

Receptacle – the expanded region at the top of the main stem (peduncle) to which the floral parts are attached. It is usually convex but may become flattened or concave.

Royal jelly – a food substance produced by nurse bees to feed to larvae destined to become queens.

Seed – the structure that develops from the fertilised ovule in seed plants (angiosperms and gymnosperms), carrying all the genetic materials to form a new diploid plant (> 'diploid').

Seed plants – seed producing plants – angiosperms and gymnosperms (> 'angiosperms', 'gymnosperms').

Sepal – individual unit of a calyx (> 'calyx').

Somatic cell – any cell of the body   i.e. any cell other than pollen, spores, gametes, or their precursors (> 'diploid').

Spadix – a specialised inflorescence in which the flowers are sessile and borne on a large fleshy axis, often with a sterile part extending beyond the inflorescence.

Spathe – a large bract enclosing a spadix (> 'bract').

Sporogenous – spore or pollen forming.

Sporophyte/sporophytic – pertaining to the haploid spore or pollen grain (> 'diploid').

Sporopollenin – the material comprising the tough outer pollen wall of most flowering plants. It comprises carbon, hydrogen and oxygen in an approximate ratio of 4:6:1. Recent results confirm the presence of fatty, aromatic and minimal carboxylic acid components. Apparently, although the components are consistent, the ratio of the components is not consistent through all plant groups. It is suggested that sporopollenin is probably 'a randomly cross-linked biomacromolecule without a repetitive large-scale structure' and, furthermore, that this is 'a characteristic which would inherently make this material resistant to enzymic attack, as well as to many laboratory procedures designed to reduce/return it to its principal components.' This being so would account for the extraordinary preservational qualities of pollen exine.

Style – the region of the pistil (carpel) between the stigma and the ovary (> 'carpel', 'ovary', 'pistil', 'stigma').

Stylar canal – a central channel through the style connecting the stigma to the ovary locule. The pollen tube grows down through the stylar canal. However, not all styles have a canal; in some species the pollen tube penetrates the style through stylar tissue (> 'style', 'stigma', 'pollen tube').

Stamens – the male pollen producing reproductive organ of a flowering plant, collectively the androecium (> 'androecium', 'pollen').

Stigma – the receptive, often highly modified, region at the top of the pistil (carpel) which receives the pollen (> 'carpel', 'pistil').

Symbiotic – an intimate relationship; in its narrow sense, of mutual benefit to both the organisms involved. However, in a wider sense it also covers relationships which can be destructive to a participant, for example, parasitism, or only of beneficial gain to one participant (commensalism), for example, epiphytic orchids.

Taxonomy – the study of the principles and practices of classification. Strictly it is applied to the study and description of variation in the natural world and the subsequent compilation of classifications (> 'families of plants').

Tetrad – a general term for a group of four united pollen grains or spores, either as a dispersal unit or as a developmental stage. There are a number of different types of pollen tetrads or dispersal units of which 'tetragonal' and 'tetrahedral' are the most frequent (> 'pollinium', 'polyad').

Vegetative cell – the largest of the two or three cells in the young pollen grain; its role is not completely understood but it is thought to be involved in the growth and development of the pollen tube (> 'generative cell').

Zygomorphic – with bi-lateral symmetry

Zygote – the product of the fusion of two gametes, before it has undergone mitosis or meiosis.

BIBLIOGRAPHY

POLLEN & BOTANY

Bailey, J. (editor) (1999) *The Penguin Dictionary of Plant Sciences.* Second edition (completely revised). Penguin Books, London, England.

Blunt, W. (1971) *The Compleat Naturalist: a Life of Linnaeus.* Collins, London.

Camus, J.M., Jermy, A.C. & Thomas, B.A. (1991) *A World of Ferns.* Natural History Museum Publications, London.

Church, A.H. (1908) *Types of floral Mechanism. Part I Types I-XII (Jan. to April).* Oxford at the Clarendon Press.

Crane, E. (1983) *The Archaeology of Beekeeping.* Duckworth, London.

Cresti, M., Blackmore, S. & Went, J.L. van (1991) *Atlas of Sexual Reproduction in Flowering Plants.* Springer-Verlag, Vienna, New York.

Dafni, A., Hesse, M., Pacini, E. (editors) (2000) *Pollen and Pollination.* Springer-Verlag, Vienna & New York.

Day Lewis, C. (transl.) (1999) *Virgil – The Eclogues and Georgics.* Oxford University Press.

Desmond, R. (1995) *Kew, the History of the Royal Botanic Gardens.* Harvill.

Erdtman, G. (1943) *An Introduction to Pollen Analysis.* Waltham Mass., USA (1954).

Erdtman, G. (1952) *Pollen morphology and plant taxonomy. Angiosperms.* Almqvist & Wiksell, Stockholm.

Erdtman, G. (1969) *Handbook of Palynology: an introduction to the study of pollen grains and spores.* Munksgaard, Copenhagen.

Faegri, K., Pijl, L. van der (1979) *The Principles of Pollination Ecology.* Third revised edition. Pergamon Press, Oxford, New York, Paris.

Fritzsche, J. (1837) *Über den Pollen.* Academie der Wissenschaften, St Petersburg.

Goodman, L. (2003) *Form and Function in the Honey Bee.* International Bee Research Association, Cardiff.

Grew, N. (1682) *The Anatomy of Flowers, prosecuted With the bare Eye, And the Microscope.* W. Rawlins, London.

Heywood, V.H. (editor) (1978) *Flowering Plants of the World.* Oxford University Press, Oxford, London, Melbourne.

Hodges, D. (1952) *The Pollen Loads of the Honey Bee.* Bee Research Association, London.

Hooke, R. (1665) *Micrographia: or some physiological descriptions of minute bodies made by magnifying glasses. With observations and enquiries thereupon.* London, printed by Jo. Martyn, and Ja. Allestry, Printers to the Royal Society.

Jardine, L. (2004) *The Curious Life of Robert Hooke,* Harper Collins, London.

Jeffrey, C. (1989) *Biological Nomenclature.* 3rd Edition. Systematics Association, Edward Arnold.

Kerner von Marilaun, A. & Oliver, F.W. (1903) *The Natural History of Plants.* Vol. II. The Gresham Publishing Company, London.

Kirk, W. (1994) *A Colour Guide to the Pollen Loads of the Honey Bee.* International Bee Research Association, Cardiff.

Knox, R.B. (1979) *Pollen and Allergy.* Institute of Biology, Studies in Biology No. 107. Edward Arnold.

Lawrence, G.H.M. (1955) *An Introduction to Plant Taxonomy.* The Macmillan Company, New York.

Lewis, D. (1979) *Sexual Incompatibility in Plants.* Institute of Biology, Studies in Biology No. 110. Edward Arnold.

Linnaeus, C. (1735) *Systema Naturae.* Leyden. (Facsimile, Stockholm 1960).

Linnaeus, C. (1746-7) *Sponsalia Plantarum.* Linnaeus President. Dissertation by Johan Gustav Wahlbom. Stockholm.

Linnaeus, C. (1750-1) *Philosophia Botanica.* Stockholm.

Moore, P.D., Webb, J.A., Collinson, M.E. (1991) *An Illustrated Guide to Pollen Analysis.* Blackwell Scientific Publications Ltd.

Melzer, W. (1989) *Beekeeping: a Complete Owner's Manual.* Barrons Educational Series Inc., New York.

Mueller, B. (1952) *Goethe's Botanical Writings.* University of Hawaii Press, Honolulu, Hawaii.

Nilsson, S. & Praglowski, J. [eds.] (1992) *Erdtman's Handbook of Palynology.* 2nd Edition. Munksgaard, Copenhagen. [A revised edition of Erdtman's 1969 handbook]

Punt, W., Blackmore, S., Nilsson, S., Le Thomas, A. (1994) *Glossary of Pollen and Spore Terminology.* LPP Foundation, Utrecht. [see also website: http://www.bio.uu.nl/~palaeo/glossary/]

Proctor, M., Yeo, P. (1973) *The Pollination of Flowers.* The New Naturalist Series, Collins, London.

Proctor, M., Yeo, P., Lack, A. (1996) *The Natural History of Pollination.* The New Naturalist Series, HarperCollins, London.

Sawyer, R. (1981) *Pollen Identification for Beekeepers.* University College Cardiff Press.

Stanley, R.G., Linskens, H.F. (1974) *Pollen: Biology, Biochemistry, Management.* Springer, Berlin, Heidelberg, New York.

Stearn, W.T. (1992) *Botanical Latin.* Fourth Edition. David & Charles, Devon.

Wodehouse, R.P. (1935) *Pollen Grains – their structure, identification and significance in science and medicine.* McGraw-Hill Book Company, Inc., New York and London.

ART

Adam, H.C. (1999) *Karl Blossfeldt.* Prestel, Munich.

Arnold, K. (2002) *Science and art: Symbiosis or just good friends?* Wellcome Trust News Supplement, London.

Asherby, J. (1996) *Mapplethorpe, Pistils.* Jonathan Cape, London.

Bataille, G. & Mattenklott, G. (1999) *Karl Blossfeldt, Art Forms in Nature.* Schirmer Art Books, Munich

Becher, B. & Becher H. (1993) *Gas Tanks.* MIT Press, Cambridge, MA.

Benke, B. (2000) *O'Keefe.* Taschen, Koln.

Blunt, W. (1950) *The Art of Botanical Illustration.* New Naturalist Series, Collins, London.

Bouquert, C. (1996) *Laure Albin Guillot.* Marval, Paris.

Cragg, A. (1998) *Anthony Cragg, Material, Object, Form.* Hatje Cantz Verlag, Ostfildern.

Darwin, E. (1791) *The Botanic Garden, A Poem in Two Parts with Philosophical Notes.* J. Nichols, London.

Davies, P.H. (2002) *Photographing Flowers and Plants.* Collins & Brown.

Dresser, C. (1876) *Studies in Design.* Studio Editions, 1988, London

Durant, S. (1993) *Christopher Dresser.* Academy Editions, London

Ede, S. (2000) *Strange and Charmed – Science and the Contemporary Visual Arts.* Calouste Gulbenkian Foundation, London.

Ewing, W. (1991) *Flora Photographica.* Thames & Hudson, London

Frankel, F. (2002) *Envisioning Science, The Design and Craft of the Science Image.* MIT Press, Cambridge, MA.

Gamwell, L. (2002) *Exploring the Invisible, Art Science and the Spiritual.* Princeton University Press, Princeton, NJ.

Haeckel, E. (1904) *Art Forms in Nature.* Reprinted, 1998. Prestel, Munich.

Herzog, H. (1996) *The Art of the Flower.* Edition Stemmele AG, Zurich.

Hewison, R. (1976) *John Ruskin, the argument of the eye.* Thames & Hudson, London.

Jones, O. (1856) *The Grammar of Ornament.* Studio Editions 1988. London.

Kemp, M. (2000) *Visualizations, the Nature Book of Art and Science.* Oxford University Press, Oxford.

Kesseler, R. (2001) *Pollinate.* Grizedale Arts and The Wordsworth Trust, Cumbria.

Mabberley, D. (2000) *Arthur Harry Church, the Anatomy of Flowers.* Merrell, 2000, London.

Moore, A. & Garibaldi, C. (2003) *Flower Power, the Meaning of Flowers in Art.* Philip Wilson, London.

Rugoff, R. & Corrin, L. (2000) *The Greenhouse Effect.* Serpentine Gallery, Catalogue, London.

Segal, S. (1990) *Flowers and Nature, Netherlandish Flower Painting of Four Centuries.* Hijnk International, Amstelveen.

Stafford, B.M. (1994) *Artful Science, Enlightenment, Entertainment and the Eclipse of the Visual Image.* MIT Press, Cambridge MA.

Stafford, B.M. (1996) *Good Looking, Essays on the Virtue of Images.* MIT Press, Cambridge MA.

Thomas, A. (1997) *The Beauty of Another Order, Photography in Science.* Yale University Press, New Haven.

Walter Lack, H. (2001) *Garden of Eden.* Taschen, Koln.

Wilde, A. & J. (2001) *Karl Blossfeldt, Working Collages.* MIT Press, Cambridge MA.

Woof, P. & Harley, M.M. (2002) *The Wordsworths and the Daffodils.* Wordsworth Trust, Cumbria.

INDEX OF PLANTS ILLUSTRATED

Liriodendron tulipifera – Tulip Tree (Magnoliaceae) –
tulip-like hermaphrodite flower